How
Animals Talk

How Animals Talk

WILLIAM J. LONG

With a new Introduction by
WILLIAM YOUNG

DOVER PUBLICATIONS, INC.
Mineola, New York

Bibliographical Note

This Dover edition, first published in 2009, is a slightly altered republication of
the work first published by Harper & Brothers, New York and London, in 1919.
The eight color plates in the original work have been omitted.

Library of Congress Cataloging-in-Publication Data

Long, William J. (William Joseph), 1867–1952.
    How animals talk / William J. Long ; with a new Introduction by William
Young.
       p. cm.
    "Slightly altered republication of the work first published by Harper &
Brothers, New York, in 1919."
    ISBN-13: 978-0-486-46880-8
    ISBN-10: 0-486-46880-1
    1. Animal communication. 2. Forest animals—Behavior. I. Title.

QL776.L63 2009
591.509—dc22

                                       2008044584

Manufactured in the United States of America
Dover Publications, Inc., 31 East 2nd Street, Mineola, N.Y. 11501

Contents

Introduction to the Dover Edition
by William Young

In 1920, Hugh Lofting published the first in a series of a dozen popular children's books about Doctor John Dolittle, who had the special talent of being able to talk with animals in their own languages. A year before the first Doctor Dolittle book was published and many years before the character became much more famous from movie portrayals by Rex Harrison and Eddie Murphy, Doctor William Joseph Long published *How Animals Talk*, a book in which he speculated about animal communication. Some of Long's work was very controversial during his lifetime, yet when reading *How Animals Talk* in the twenty-first century, one can appreciate him as an astute and sensitive observer of nature.

Long was born in 1857, two years before Darwin published *On the Origin of Species*, and died in 1952, a year before James Watson and Francis Crick published their theory about the double helix structure of DNA. The views he expressed in *How Animals Talk* reflect a strong belief in the similarities between humans and other animals which were postulated by Darwin and confirmed through the work of Watson, Crick, and others who have studied the chemical makeup of life. Long wrote, "At the bottom of it, I suppose, is the fact that in every wild or natural creature is something, at once mysterious and familiar, which appeals powerfully to your interest or sympathy, as if you saw a faint shadow of your other self, or caught a fleeting memory of that vanished time when you lived in a child's world of wonder and delight."

Biologists and naturalists face a difficult challenge when trying to imagine how non-human animals think and communicate. Because there is no common spoken language between humans and non-humans, people have no way of discussing the subject with non-humans. Any statements humans make about non-human thought processes must be based on observations or hypotheses. Complicating the issue is the deep and long-standing prejudices among people inside and outside the scientific community about humans being the only creatures capable of various thought processes and cultural activities. The scientific community can be extremely rigid and dogmatic toward anyone who proposes theories that differ significantly from what is commonly accepted. Among the theories that scientists have rejected in the past and, in some instances, continue to reject are whether humans are the only animals capable of using tools, whether non-human animals are capable of having "culture," and whether non-humans are acting entirely by instinct rather than being capable of learning.

Scientists who cling to prejudices about the limited abilities of non-human minds typically level the charge of "anthropomorphism" against anyone who suggests that other animals are engaging in thought processes or behaviors similar to humans. In the introduction to his 1996 book *The Minds of Birds*, the tropical naturalist Alexander Skutch questioned why anthropomorphism is disparaged by biologists who talk about the minds of animals, even though the practice is considered acceptable when comparing the anatomy of humans and non-humans. Skutch suggested that anatomical similarities between humans and non-humans are among the strongest evidence to support the theory of evolution, so it is reasonable to think that animals might in some measure resemble humans psychically. When observing animals, Long clearly was not bothered by accusations of anthropomorphism. He saw himself in them and wondered how closely they resembled him. Rather than viewing non-humans as inferiors, he viewed them as different, having abilities superior to humans for some tasks and inferior to humans for others. He was fascinated by the superior sensory abilities of some creatures, which is why he thought that the simple act of taking a dog for a walk could be a revelation for anyone who could do it without prejudices about the way the dog is supposed to be.

Long devotes a chapter of *How Animals Talk* to the concept of *chumfo*, which he defines as the extraordinary sensory powers of animals. He borrowed the word from an African tribe living near Lake Mweru, which today is on the border of Zambia and the Democratic Republic of Congo. Long believed that living creatures using their senses must be the basis of nature study or else it would be merely "a thing of books or museums or stuffed skins or Latin names, from which all living interest has departed." Long equated nature with life, and life with strong and deep sensory perceptions of one's environment. The "thrill and mystery of awakening life" is the aspect of the natural world that excited Long the most. He believed that the natural animal or the natural man is wholly alive and awake, and that every cell has the faculty of perception. He thought that our human senses have unused possibilities and that humans might possess extra senses of which we are not conscious.

Long was a minister in the First Congregationalist Church in Stamford, Connecticut. The tone of his writing sometimes reflects the influence of fellow New Englanders Ralph Waldo Emerson and Henry David Thoreau. He earned his doctoral degree from the prestigious University of Heidelberg in Germany. Long wrote dozens of books, most of which were about nature, but he also wrote detailed texts about English and American literature, of which he had a virtually encyclopedic knowledge. At times, Long's writing blurs the line between what he believed to be true and what he knew to be true—a stylistic problem one might expect from someone accustomed to speaking from a pulpit rather than a dais at a scientific conference.

One of the six chapters in the first section of *How Animals Talk* is devoted to "Natural Telepathy." The term "telepathy" immediately raises hackles in the scientific community because of associations with the occult. Yet observers of nature in the first quarter of the twentieth century must have been justifiably puzzled and amazed when trying to figure out some of the seemingly miraculous but inscrutable communication they observed. Long spent a lot of time observing swarms of honeybees without having access to the research about honeybee communication that subsequently led to Karl von Frisch winning a Nobel Prize in medicine in 1973. Likewise, he did not know about Katy Payne's research in the 1980s and 1990s which revealed that the seemingly telepathic communication among elephants was based on the use of sounds at a frequency below the level of human hearing. Long tried to learn about animal senses from what he perceived from his own.

The second section of the book is called "How to Know the Wood Folk" and might be of practical use to anyone interested in studying birds, mammals, or other forms of wildlife. It provides detailed techniques for trying to get physically close to creatures. Long believed that wild creatures are not normally governed by fear and terror; he thought fear was an exclusively human affliction, "as artificial as sin, which the animal escapes by virtue of being natural." He stressed that animals do not like to be stared at, but that they have a lively interest in every new thing they meet. Hence, the fact that you are not one of them does not necessarily mean that they will run away from you. He said that people who are clean and not sweating profusely are less likely to give off a scent that animals will detect. He also suggested that people wear neutral-colored clothes that harmonize with the colors of the woods.

When leading bird watching trips, I have had to deal with the difficulty of trying to show people birds while there is a lot of talking and movement in the group. Motion and noise tend to frighten wildlife. Long believed in the importance of being physically and mentally still when observing animals. The mental stillness is the most difficult aspect to explain to people not experienced in field observation. Having an unfocused, wandering mind will inhibit your ability to view wildlife. Some of Long's ideas about field observation have a flavor of Eastern religions, with the body and mind achieving a stillness and focus that become one with their surroundings. He believed that the best way to see wildlife was to sit still in one place and let inquisitive animals come to him rather than roaming noisily through the woods in search of them.

The third and final section of the book contains Long's reminiscences about a pond and the wildlife around it. This section contains some of his most spiritual reflections about nature. Like Thoreau, Long cherished solitude and never felt lonely when he was alone in the woods. He thought that "ideals may be quite as companionable as folks" and that "around you in a goodly company are beauty, peace, spacious freedom and harmonious thoughts." In the ongoing battle between humans and nature, Long clearly stakes out a strong position in support of the latter, believing that nature is constantly trying to rebuild the environmental wonders that humans destroy. Were he alive today, he would most likely be speaking out strongly for environmental protection.

Americans in the twenty-first century who read *How Animals Talk* will have trouble understanding why some of Long's previous books were so controversial that the president of the United States publicly denounced them. Yet President Theodore Roosevelt attacked some of Long's stories for their lack of veracity. The famous naturalist John Burroughs, a friend of Roosevelt, wrote a long article in a 1903 issue of *The Atlantic Monthly* in which he echoed Roosevelt's position, making the distinction between writers of "real and sham natural history." Burroughs and Roosevelt accused Long of being a "nature faker." In *How Animals Talk*, Long took a swipe at Roosevelt, an avid hunter, when he wrote: ". . . you can learn nothing worth knowing about birds or beasts so long as you seek them with a gun in your hand. On that road you shall find only common dust, and at the end of it a valley of dry bones. Whether you carry the gun frankly for sport, or delude yourself with the notion that you can add to natural history by collecting more skins or skulls, you have unconsciously placed destruction above fulfillment, stark death above the beautiful mystery of life. So must you estrange both the animal and yourself, making it impossible for you to meet on any common ground of understanding."* Roosevelt died at the beginning of 1919, the year *How Animals Talk* was published.

---

* Long was not averse to shooting animals; he simply did not think that effective field observation could be possible on a hunting expedition.

The conflict between Burroughs and Long was in some ways more of style than substance. If you read the works of both men, you will detect a similar deep love and reverence for the natural world. Their respective positions are not diametrically opposed, as would be the case between a climate change denier and someone who believes that global warming poses catastrophic environmental threats. Their disagreement was more like the difference between a monotheist who has developed a set of moral values by reading religious texts and a humanist who has developed a similar set of moral values by reading the works of secular philosophers.

The early work of Long contains some stories about nature that clearly stray far beyond the truth, although hardly in a way that would warrant intervention by the leader of the free world. Even the bullying Roosevelt would not have had the courage to similarly attack the numerous religions whose scriptures contain stories that do not seem scientifically plausible. With respect to Long's occasional nature fakery, the issue is whether a tall tale designed to encourage a love of nature does more harm than good. There are enough miracles in nature that writers such as Long did not need to invent stories to impress readers. Still, a tall tale that induces people, especially youngsters, to develop an interest in nature might end up being more beneficial than a dry scientific account that fails to arouse curiosity about nature.

In the *Atlantic Monthly* article, Burroughs attacked Long for suggesting that animals educate their young and that instinct plays a much smaller role than previously supposed. Burroughs cynically noted: "This is indeed a new suggestion in the field of natural history. What a wonder that Darwin did not find it out, or the observers before and since his time. But the honor of the discovery belongs to our own day and land!" On this point, Long showed that he was clearly ahead of his time, and Burroughs' over-emphasis of nature over nurture has been disproved by subsequent scientific research. For instance, some songbirds are born without knowing how to sing the typical song of their species. They learn their song by listening to it sung by others of their kind. Frans de Waal's 2001 book *The Ape and Sushi Master* provides a long and detailed examination of the concept of animal culture, which involves the passing of useful knowledge within and between generations of a species. A prime example is Japanese monkeys teaching each other to wash sweet potatoes before eating them. Such behavior is learned and not done by instinct.

In looking back on the "nature faker" controversy more than a century later, one realizes that science and natural history issues are a lot more complex than critics such as Roosevelt and Burroughs realized, and that some of the controversial ideas of William Long might not have been so far-fetched. What is accepted as scientific truth in one generation might be subsequently discarded by future generations, and ideas that are scorned in one generation might later be accepted in another. The scientific method consists of observation, hypothesis, and then testing to determine whether the hypothesis is true. Long spent a great deal of time observing and hypothesizing. The reissue of *How Animals Talk* will allow contemporary readers to explore and enjoy the hypotheses of an important early twentieth century naturalist. Many of the issues he explored about the sensory and mental capabilities of animals are as relevant today as when he was writing about them. And his infectious love and enthusiasm for nature does not in any way seem diminished with the passage of time.

How Animals Talk

A Little Dog-Comedy

I

Dᴵᴰ you ever see two friendly dogs meet when one tried to tell the other of something he had discovered, when they touched noses, stood for a moment in strange, silent parley, then wagged their tails with mutual understanding and hurried off together on a canine junket?

That was the little comedy which first drew my attention to the matter of animal communication, many years ago, and set my feet in the unblazed trail we are now to follow. And a very woodsy trail you shall find it, dim and solitary, with plenty of "blind" spots where one may easily go astray, and without any promise of what waits at the other end of it.

One summer afternoon I was reading by the open window, while my old setter, Don, lay flat on his side in the shade of a syringa-bush. He had scooped out a hollow to suit him, and was enjoying the touch of the cool earth when a fat little terrier, a neighbor's pet, came running with evident excitement to wake the old dog up. Don half raised his head, recognized his friend Nip and thumped the ground lazily with his tail.

"It's all right, little dog. You're always excited over something

3

of no consequence; but don't bother me this hot day," he said, in dog-talk, and dropped his head to sleep again.

But Nip was not to be put aside, having something big on his mind. He nudged Don sharply, and the old dog sprang to his feet as if galvanized. For an interval of perhaps five seconds they stood motionless, tense, their noses almost touching; then Don's plume began to wave.

"Oh, I see!" he said; and Nip's stubby tail whipped violently, as if to add, "Thank Heaven you do, at last!" The next moment they were away on the jump and disappeared round a corner of the house.

Here was comedy afoot, so I slipped out through the back door to follow it. The dogs took no notice of me, and probably had no notion that they were observed; for I took pains to keep out of sight till the play was over. Through the hay-field they led me, across the pasture lot, and over a wall at the foot of a half-cultivated hillside. Peering through a chink of the wall, I saw Nip dancing and barking at a rock-pile, and between two of the rocks was a woodchuck cornered.

For weeks Nip had been laying siege to that same wood-chuck, which had a den on the hillside in a patch of red clover, most convenient to some garden truck. A dozen times, to my knowledge, the little dog had rushed the rascal; but as Nip was fat and the chuck cunning, the chase always ended the same way, one comedian diving into the earth with a defiant whistle, leaving the other to scratch or bark impotently outside.

Any reasonable dog would soon have tired of such an uneven game; but a terrier is not a reasonable dog. At first Nip tried his best to drag Don into the affair; but the old setter had long since passed the heyday of youth, when any kind of an adventure could interest him. In the presence of grouse or woodcock he would still become splendidly animate, and then the years would slip from him as a garment; but to stupid groundhogs and all such "small deer" he was loftily indifferent. He was an aristo-crat, of true-blue blood, and I had trained him to let all crea-tures save his proper game severely alone. So, after following Nip once and finding nothing more exciting than a hole in the ground, with the familiar smell of woodchuck about it, he had left the terrier to his own amusement.

When speed failed, or wind, it was vastly amusing to watch Nip try to adopt cat-strategy, hiding, creeping, scheming to cut off the enemy's retreat. Almost every day he would have another go at the impossible; but he was too fat, too slow, too clumsy, and also too impatient after his doggy kind. By a great effort he could hold still when his game poked a cautious head out of the burrow for a look all around; but no sooner did the chuck begin to move away from his doorway than the little dog began to fidget in his hiding-place, and his tail (the one part of a dog that cannot lie) would wildly betray his emotions. Invariably he made his rush too soon, and the woodchuck whistled into his den with time to spare.

On this summer afternoon, however, Nip had better luck or used better tactics. Whether he went round the hill and came over the top from an unexpected quarter, or lay in wait in his accustomed place with more than his accustomed patience, I have no means of knowing. By some new device or turn of luck he certainly came between the game and its stronghold; whereupon the chuck scuttled down the hill and took refuge among the rocks. There Nip's courage failed him. He was a little dog with a big bark; and the sight of the grizzled veteran with back against a stone and both flanks protected probably made him realize that it is one thing to chase a chuck which runs away, but quite another thing to enter his cave while he stands facing you, his beady eyes snapping and his big teeth bare. So after a spell of brave barking Nip had rushed off to fetch a larger dog.

All that was natural enough, and very doglike; at least it so appeared to me, after seeing other little dogs play a similar part; but the amazing feature of this particular comedy was that Nip had no difficulty in getting help from a champion who had refused to be interested up to that critical moment. Through the wall I saw him lead Don straight to the rocks. The old dog thrust in his head, yelped once as he was bitten, dragged out the chuck, gave him a shake and a quieting crunch; then, without the slightest evident concern, he left Nip to worry and finish and brag over the enemy.

It is part of the fascination of watching any animal comedy that it always leaves you with a question; and the unanswerable question here was, How did Nip let the other dog know what he wanted?

If you are intimate enough with dogs to have discovered that they depend on their noses for all accurate information, that they have, as it were, a smellscape instead of a landscape forever before them, you will say at once, "Don must have smelled woodchuck"; but that is a merely convenient answer which does not explain or even consider the facts. Don already knew the general smell of woodchuck very well, and was, moreover, acquainted with the odor of the particular woodchuck to which his little dog-chum had been laying siege. He knew it at first hand from the creature itself, having once put his nose into the burrow; he got a secondary whiff of it every time Nip returned from his fruitless digging; and he was utterly indifferent to such foolish hunting. Many times before the day of reckoning arrived Nip had rushed into the yard in the same excitement, with the same reek of earth and woodchuck about him; and, so far as one may judge a dog by his action, Don took no interest in the little dog's story. Yet he was off on the instant of hearing that the familiar smell of woodchuck now meant something more than a hole in the ground.

That some kind of message passed between the two dogs is, I think, beyond a reasonable doubt; and it is precisely this silent and mysterious kind of communication (the kind that occurs when your dog comes to you when you are reading, looks intently into your face, and tells you without words that he wants a drink or that it is time for him to be put to bed) that I propose now to make clear. Before we enter that trail of silence, how-

ever, there is a much simpler language, such as is implied in the whistle of a quail or the howl of a wolf, which we must try as best we can to interpret. For unless our ears are keen enough to distinguish between the food and hunting calls of an animal, or between bob-white's love note and the yodel that brings his scattered flock together, it will be idle for us to ask what message or impulse a mother wolf sends after a running cub when she lifts her head to look at him steadily, and he checks his rush to return to her side as if she had made the murky woods echo to her assembly clamor.

Cries of the Day and Night

II

THE simplest or most obvious method of animal communication is by inarticulate cries, expressive of hunger, loneliness, anger, pleasure, and other primal needs or emotions. The wild creatures are mostly silent, and so is the bulk of their "talk," I think; but they frequently raise their voice in the morning or evening twilight, and by observing them attentively at such a time you may measure the effect of their so-called language. Thus, you see plainly that to one call the animal cocks his ear and gives answer; at another call he becomes wildly excited; a third passes over him without visible result; a fourth sets his feet in motion toward the sound or else sends him flying away from it, according to its message or import.

That animal cries have a meaning is, therefore, beyond serious doubt; but whether they have, like our simplest words, any definite or unchanging value is still a question, the probable answer being "No," since a word is the symbol of a thought or an idea; but animals live in a world of emotion, and even our human emotions are mostly dumb or inarticulate. I must give this negative answer, notwithstanding the fact that I have learned to call various birds and beasts, and that I can meet Hotspur's challenge on hearing Glendower boast that he can call spirits from the vasty deep:

> Why, so can I, or so can any man;
> But will they come when you do call for them?

Yes, the birds and beasts will surely come if you know how to give the right call; but I am still doubtful whether among themselves their audible cries are ever quite so intelligible as is their silence.

This question of animal speech has received a different and more positive answer, by the way, from a man who has spent many years in persistent observation of wild apes and monkeys. After watching the lively creatures from his cage in the jungle, attracting them by means of various fruits and recording their jabber in a phonograph, he claims to have discovered the monkey words for food, water, danger and other elementary matters. Moreover, when his phonograph repeats these simian words the monkeys of another locality seem to understand them, since they run to the proper dish at the word "food" or show evident signs of alarm at the word "danger."

It is doubtless much easier to deny such a conclusion than to prove or disprove it; but denial is commonly the first refuge of ignorance and the last of dogmatism, and with these we are not concerned. I do not know whether Garner claims too much or too little for his monkeys; I have never had opportunity to test the matter in the jungle, and the caged monkeys with which I have occasionally experimented are too debased of habit or too imbecile in their affections to interest one who has long dealt with clean wild brutes. At times, however, when I have watched a monkey with an organ-grinder, I have noticed that the unhappy little beast displays a lively interest in the chitter of chimney-swifts—a lingo which to my dull ears sounds remarkably like monkey-talk. But that is a mere impression, momentary and of little value; while Garner speaks soberly after long and immensely patient observation.

To return to first-hand evidence: among wild creatures of my acquaintance the crows come nearer than any others to some-

thing remotely akin to human speech. Several times I have
known a tame crow to learn a few of our words and, what is
much more significant, to show his superiority over parrots and
other mere mimics by using one or more of the words intelligently. There was one crow, for example, that would repeat the
word "hungry" in guttural fashion whenever he thought it was
time for him to dine. He used this word very frequently when
his dinner or supper hour drew nigh, giving me the impression,
since he did not confuse it with two other words of his vocabu-
lary, that he associated the word with the notion of food or of
eating; and if this impression be true to fact, it indicates more
than appears on the surface. We shall come to the wild crows
and their "talk" presently; the point here is, that if this bird
could use a new human word in association with a primal need,
there is nothing to prevent him from using a sound or symbol of
his own in the same way. In other words, he must have some
small faculty of language.

Another tame crow, which an imaginative boy named Pharaoh
Necho because of his hippety-hop walk, proved himself inordi-
nately fond of games, play, social gatherings of every kind. To
excitement from any source, whether bird or brute or human,
he was as responsive as a weather-vane; but his play ran mostly
to mischief, or to something that looked like joking, since he
could never see a contemplative cat or a litter of sleepy little
pigs without going out of his way to tweak a tail and stir up
trouble. At times he would watch, keeping out of sight in a leafy
tree or on the roof of the veranda, till Tabby, the house cat,
came out and sat looking over the yard, her tail stretched out
behind her. If she lay down to sleep, or sat with tail curled
snugly around her forepaws, she was never molested; but the
moment her tail was out of her sight and mind Necho saw the
chance for which he had apparently been waiting. Gliding
noiselessly down behind the unconscious cat he would tiptoe up
and hammer the projecting tail with his beak. It was a startling
blow, and at the loud squall or spitting jump that followed he
would fly off, "chuckling" immoderately.

When Necho saw or heard a gang of boys assembled he
would neglect even his dinner to join them; and presently, with-

out ever having been taught, he announced himself master of a new art by yelling, "*Ya-hoo! Come on!*" which was the rallying-cry of the clan in that neighborhood. He said this in ludicrous fashion, but unmistakably to those who knew him. Sometimes he would croak the words softly to himself, as if memorizing them or pleased at the sound; but for the most part he waited till boys were gathering for a swim or a ball game, when he would launch himself into flight and go skimming down the road, whooping out his new cry exultantly. What meaning he attached to the words, whether of boys or fun or mere excitement, I have no means of knowing.

After learning this much of our speech Necho took to the wild, following a call of the blood, I think; for it was springtime when he disappeared, and the crows' mating clamor sounded from every woodland. These birds are said to kill every member of their tribe who returns to them after living with men, and the saying may have some truth in it. I have noticed that many tame crows are like tame baboons in that they seem mortally afraid of their wild kinsmen; but Necho was apparently an exception. If he had any trouble when first he returned to his flock, the matter was settled without our knowledge, and during the following autumn there was evidence that he was again in good standing. Long afterward, as I roamed the woods, I might hear his lusty "*Ya-hoo! Come on!*" from where he led a yelling rabble of crows to chivvy a sleeping owl or jeer at a running fox; and occasionally his guttural cry sounded over the tree-tops when I could not see him or know what mischief was afoot. He never returned to the house, and never again joined our play or allowed a boy to come near him.

Not all crows have this "gift of speech"; and the fact that one tame crow learns to use a few English words, while five or six others hold fast to their own lingo, has led to the curious belief that, if you want to make a crow talk, you must split his tongue. How such a belief originated is a mystery; but it was so fixed and so widespread when I was a boy that no sooner was a young crow taken from a nest than jack-knives were sharpened, and the leathery end of the crow's tongue was solemnly split after grave debate whether a seventh or a third part was the proper medicine. If the crow talked after that, it was proof positive that the belief was true; and if he remained dumb, it was a sign that there was something wrong in the splitting; which is characteristic of a large part of our natural-history reasoning. The debates I have heard or read on the "unanswerable" question of how a chipmunk digs a hole without leaving any earth about the entrance (a question with the simplest kind of an answer) are mostly suggestive of the split-tongue superstition of crow language.

Of the tame crows I have chanced to observe, only a small proportion showed any tendency to repeat words; and these gifted ones are, I judge, the same crows that in a wild state may occasionally be heard whistling like a jay, or "barking" or "hooting" or making some other call which ordinary crows do not or cannot make, and which shows an individual talent of mimicry. This last, which I have repeatedly observed among wild crows, is a very different matter from speech; but from the fact that these mimics learn to use a few English words more or less intelligently one might not be far wrong in concluding that every crow has in his brain a small undeveloped nest of cells corresponding to our "bump" of language.

A closer observation of the wild birds may confirm this possibility. Thus, when you hear a solitary crow in a tree-top crying, *"Haw! Haw!"* monotonously, dipping his head or flirting his tail every time he repeats it, you may be sure that somewhere within range of his eye or voice a flock of his own kind are on the ground, feeding. That this particular *haw* is a communication to his fellows, telling them that the sentinel is on watch and all is well, seems to me very probable. There are naturalists, I know, who ingeniously resolve the whole phenomenon into blind chance or accident; but that does not square very well with the intelligence of crow nature as I have observed it; nor does it explain the fact that once, when I avoided the sentinel and crept near enough to shoot two members of the flock he was supposedly guarding, the rest were no sooner out of danger than they whirled upon the recreant and beat him savagely to the ground.

If you are interested enough to approach any crow-sentinel in a casual or indifferent kind of way (he will take alarm if you approach quickly or directly), you must note that his *haw* changes perceptibly while you are yet far off. It is no longer formal or monotonous; nor is it uttered with the same bodily attitude, as your eyes plainly see. You would pronounce and spell the cry exactly as before (it should be written *aw* or *haw*, not *caw*, for there is no consonant sound in it); but if your ears are keen, they will detect an entirely different accent or inflection, as they detect different accents and meanings when a sailor's casual or vibrant "Sail ho!" sings down from the crow-nest of a ship. Now run a few steps toward the sentinel, or pretend to hide and creep, and instantly the *haw* changes again. This time the accent is sharper even to your dull ears; and hardly is the cry uttered when all the crows of the unseen flock whirl into sight, heading swiftly away to the woods and safety.

Apparently, therefore, this simple *haw* of the crow is like a root word of certain ancient languages, the Chinese, for example, which has several different intonations to express different ideas, but which all sound alike to foreign ears, and which are spelled alike when they appear in foreign print. To judge by the crows' action, it is certain that their elementary *haw* has at least three distinct accents to express as many different meanings: one of "all's well," another of "watch out," and a third of "be off!" Moreover, the birds seem to understand these different meanings as clearly as we understand plain English; they feed quietly while *haw* means one thing, or spring aloft when it means another; and though you watch them a lifetime you will see nothing to indicate that there is any doubt or confusion in their minds as to the sentinel's message.

Not only the crows, but the wild ducks as well, and the deer and the fox and many other creatures, seem to understand crow-talk perfectly, or at least a part of it which concerns their own welfare. Thus, on the seacoast in winter you hear the crows *hawing* continually as they follow the tide-line in search of food. For hours this talk goes on, loudly or sleepily, and the wild ducks pay absolutely no attention to it; though they must know well that hungry crows will kill a wounded or careless duck and eat him to the bones whenever they have a chance. Because of this dangerous propensity you would naturally expect the water-fowl to be suspicious of the black freebooter and to be alert when they see or hear him; but no sooner do you begin to hunt with a gun than you learn a thing to make you respect the crow, and perhaps to make you wonder how much or how very little you know of the ways of the wood folk.

Many of the ducks, the black or dusky mallards especially, like to come ashore every day in a secluded spot under the lee of a bank, there to rest or preen or take a quiet nap in company. It is a tempting sight to see a score or a hundred of the splendid birds in a close group, their heads mostly tucked under their wings; but it is practically impossible to stalk them, for the reason that the crows are forever ranging the shore, and a crow never passes a group of sleeping ducks without lifting his flight to take a look over the bank behind them. What his motive is no man can say; we only note that, in effect, he stands sentinel for the ducks against a common enemy, as he habitually does for his own kind. There is no escaping that keen, searching glance of his; he sees you creeping through the beach-grass or hiding behind a bush. He flings out a single *haw!* with warning, danger, derision in it; and now the same ducks that have heard him all day without concern spring aloft on the instant and head swiftly out to sea.

The crows have several other variations of the same cry, expressive of other matters, which all the tribe seem to understand clearly, but which are meaningless to human ears. When I imitate the distress-call of a young crow, for example, I can bring a flock over my head at almost any time, the only condition being that I keep well concealed. At the first glimpse of a man in hiding they sheer off, and it is seldom that I can bring them back a second time to the same spot; yet I have a companion, one who utters a call very much like mine to ordinary ears, who can bring the flock back to him even after they have seen him and suffered at his hands. More than once I have stood beside him in the woods and fired a gun repeatedly, killing a crow and scattering the flock pell-mell at every shot; but no sooner does he begin to talk crow-talk than back they come again. What he says to them that I do not or cannot say is something that only the crows understand.

It is commonly assumed that they come to such a call because they hear in it a cry for help from one of their own kind. That is undoubtedly true at times; for a help-call, especially from a cub or nestling, is a summons to which most animals and birds instinctively respond. And, strangely enough, the smaller they are the braver they seem to be. A mother-partridge has more than once flown in my face or beaten me with her wings, while "fierce" hawks, owls and eagles have merely circled around me at a safe distance when I came near their young. In the majority of cases, however, I think that birds come to a distress-call simply because the excitement of an individual spreads to all creatures within sight or hearing, just as a crowd of men or women will become excited and rush to a common center before they know what the stir is all about.

In confirmation of this theory, it is not necessary to cry like a distressed young crow to bring a flock over your head. The imitated *hawing* of an old crow will do quite as well, if you throw the proper excitement into it. Again, on any summer day you will hear in your own yard the *pip-pip* of arriving or departing robins. The same call is uttered by both sexes, at all times and in all places; yet if you listen closely you must note that there is immense variety in the accent or inflection of even this simple sound. The call is clear, ringing, joyous when the robins first arrive in the spring; it is subdued when they gather for the autumn flight; it is sleepy or querulous when they stand full-fed by the nest, and most business-like when they launch themselves into flight, which is the moment when you are most sure to hear it. A robin utters this call hundreds of times every day, in one accent or another, and neither the other robins nor their feathered neighbors seem to pay any attention to it; but when a red squirrel comes plundering a nest, and the mother robin sends forth the same *pip-pip* with a different intonation, then the response is instantaneous. The alarm spreads swiftly over wood and field; clamor uprises, and birds of many species come rushing in from all directions; not because they have heard that Meeko is again killing young robins (at least, it does not seem so to me), but because excitement is afoot, and they are bound to join it or find out about it before they can settle down comfortably to their own affairs.

There is an interesting way by which you may test this contagion of excitement for yourself. Hide at the edge of the woods or in any other bird neighborhood in the early morning, preferably at a season when every nest has eggs or fledglings in it; press two fingers against your lips and draw the breath sharply between them, repeating the squeaky cry as rapidly as possible. The sound has a peculiarly exciting quality even to human ears (twice have I seen men run wildly to answer it), and birds come to it as boys to a fire alarm. In a few moments you may have them streaming in from the four quarters of bird world, all highly excited, and perhaps all ready to protect some innocent nest from snake or crow or squirrel. Because the response is most electric at the season when fledglings are most helpless, you are apt to think that this call of yours is mistaken by mother birds for a cry for help. That may be true; but be not too sure about it. The fledglings themselves will come almost as readily to the call when the nesting season is over and gone.

I have tried that same exciting summons in many places, wild or settled, and commonly but not invariably with the same result, as if it were a word from the universal bird language. Once in a secluded valley of northern Italy I saw a hunter with his gun, and promptly forgot my own errand in order to chum with him and find out what he had learned of the wood folk. He was hunting birds to eat. "Those birds there!" he said, pointing to a passing flock which I did not recognize, but which seemed pitifully small game to me. Presently I learned that he could not shoot flying, and was having such bad luck that, he said, the devil surely had a hand in it. He was a smiling, companionable loafer, and for a time I tagged after him, watching him amusedly as he made careful but vain stalks of little birds that seemed to have been made wild by much hunting. In a spirit of thoughtless curiosity, and perhaps also to test bird nature in a strange land, I invited the hunter to hide with me in a thicket while I gave the call which had so often brought the feathered folk of my own New England woods. At my cry a wisp of birds whirled in to light at the edge of the covert; the

Italian's gun roared; and then I discovered that the wretch was killing skylarks.

I have since had many an uncomfortable moment at the thought of how many lovely songsters may have paid with their lives for that ungodly experiment; for my companion hailed me as a master Nimrod from the New World; and when I refused, on the plea of bad luck, to teach him the call, I heard him give a distressingly good imitation of it. Yet the experiment seemed to prove that everywhere birds quickly catch the contagion of excitement; that in many cases they respond to a call because it stirs their anger or curiosity rather than because it conveys any definite summons for help or warning of danger.

When you open your ears among the beasts you hear precisely the same story; that is, certain cries apparently have definite meaning, like the accented *haw* of a crow, while others convey and also spread a wild emotion. Of all beasts, the wolves are perhaps the keenest, the most intelligent, and these seem to have definite calls for food or help or hunting or assembly. Such calls are strictly tribal, I think, like the dialects of Indians, since the call of a coyote is quite different from the call of a timber wolf even when both intend to convey the same meaning. A friend of mine, an excellent mimic, who spent many years in the West, has shot more than a score of coyotes after drawing them within range by sending forth the food-call in winter; but though he knows also the food-call of the timber wolf, he has never once deceived these larger brutes by his imitation of it; nor has he ever seen a wolf of one species respond to the food or hunting call of another.

Like most other wild animals, timid or savage, the sensitive wolves all respond, but much more warily than the birds, to almost any inarticulate cry expressive of emotional excitement; just as your dog, who is yesterday's wolf, grows uneasy when you whine in your nose like a distressed puppy, or leaps up, ready to fly out of door or window, when a wild *ki-yi* breaks out in the distance. Indeed, it is easier to keep a boy from a fire than a dog from a crowd or excitement of any kind; and the same is true of their wild relatives, though the wariness of the latter keeps them hidden where you cannot follow their action. The greatest commotion I ever witnessed in a timber-wolf pack was occasioned by the moaning howl of a wounded wolf on a frozen lake in midwinter. It was a cry utterly unlike anything I had ever recorded up to that time, and every time they heard it the grim beasts ran wildly here and there, howling like lunatics. Then, when the wounded one grew quiet, they would approach and sniff him all over; after which some would sit on their tails and watch him closely, while others circled about on the ice, using their noses like hounds in search of a lost trail.

Occasionally, when I have had these uncanny brutes near me in the North, I have tried to call them or make them answer by giving what seemed to me a very good imitation of their cries; but seldom has a howl of mine been returned. On the contrary, the brutes almost always stop their howling whenever I begin to talk wolf-talk, as if they were listening and saying, "What under the moon is that now?" Then old Tomah, the Indian, comes out of his blanket and gives a howl exactly like mine, but with something in it which I cannot fathom or master, and instantly from the snow-filled woods comes back the wild wolf answer.

Likewise, I have called moose in many different localities, and am persuaded that it makes very little difference what kind of whine or grunt or bellow you utter, since anything resembling a moose-call will do the trick if you know how to put the proper feeling into your voice. After listening carefully to many callers, I note this characteristic difference: that one man invariably makes the game wary, suspicious, fearful, no matter how finely he calls; while another in the same place, with the same trumpet and apparently with the same call, manages to put something into his voice, something primal, emotional and essentially *animal*, which brings a bull moose hurriedly to investigate. Thus it happens that the worst caller I ever heard—worst in that he had no sense, no cunning, no knowledge of moose habits, and uttered a blatant, monstrous roar unlike anything a sane man ever heard in the heavens above or the earth beneath—was still the most successful in getting his game into the open. Three nights in succession I heard him call in a region where moose were over-shy from much hunting, and where my own imitation of the animal's natural voice brought small response. In that time fourteen bulls answered him, all that were within hearing, I think; and every one of the great brutes threw caution to the dogs and came out on the jump.

From such observations, and from others which I have not chronicled, I judge that the higher orders of birds and beasts have a few calls which stand for definite things, or mental images of things, but that their ordinary cries merely project an emotion or excitement in such a way that it stirs a similar emotion in other birds or beasts of the same species; just as the sound of hearty laughter invariably stirs the feeling of mirth in men who hear it, or any inarticulate cry of fear sets human feet in motion—toward the cry if the hearer be brave, or away from it if he be of cowardly disposition. Yet even among men, who by civilization have lost some of their natural virtues, the primal impulse still lives. Like the wolf or the raccoon, the man's first impulse is to rush to his distressed or excited fellows. If he turns and runs the other way, it means simply that his artificial habit or training has deadened his natural instincts.

In speaking of "man" here I refer to the genus *homo,* not to the male specimen thereof. Among brutes most of the natural instincts are the same in both sexes; they vary in degree, not in kind, and the instincts of the female are commonly the stronger or keener. Yet I have noticed, or think I have noticed, this difference: when a cry of distress is uttered in the woods, the first bird or beast to appear is almost always a female; but the male is quicker on his toes at a battle-yell or a senseless clamor.

This last is a personal impression, and cannot well be verified. The only record I have which might pass for evidence in the matter comes from my observation of the crows. In the spring many of these questionable birds indulge their taste for eggs or tender flesh and soon become incurable nest-robbers; and for that reason I often shoot them, to save other and more useful birds. The method is very simple: one hides and calls, and takes the crows as they appear in swift flight, the number shot being commonly limited to one or two at a time. And I have observed repeatedly, at different times and in different localities, that when I use the distress-call of a young crow as a decoy, the first to appear over the tree-tops is a female. This is the common rule, with occasional exceptions to point or emphasize it. But whenever I clamor like a crow that has discovered an owl, or send forth a senselessly excited *hawing,* almost invariably the first crow to come whooping over is a long-winged and glossy old male.

Does it seem to you like thoughtless barbarity on my part to kill crows in this fashion? Perhaps it is barbarous; I do not quite know; but it certainly is not thoughtless. One cannot blame the crows for their taste in eggs or nestlings; but one must note that they destroy an enormous number of insectivorous birds, and that the harm they do in this respect outweighs their usefulness in destroying field-mice and beetles. I write this with regret; for I admire the crow, and consider him as, of all birds, the most intelligent and the most considerate of his own kind. I know that it is a moot question whether the crow does more harm or good, and that some naturalists have settled it in his favor; but I have too often caught him plundering nests in the springtime to be much impressed by his alleged usefulness at other seasons. I think that he may have been once useful in preserving the so-called balance of nature; but that balance is now dangerously unequal. The crow has flourished even in well-settled regions, thanks to his superior wit, while other useful birds have fearfully diminished, and this at a time when our orchards and gardens call more and more insistently for their help. Because of his disproportionate numbers the crow now appears to me, like our destructive and useless cats, as a positive menace in a country where he once occupied a modest or inconspicuous place—such a place as he still occupies in the wilderness, where I meet him but rarely, and where I am glad to leave him in peace, since he does not seriously interfere with his more beautiful or more useful neighbors. But we are wandering from the dim trail of animal communication, which we set out to follow.

The inarticulate but variously accented cries of which we have spoken constitute the only animal language to which our naturalists have thus far paid any attention; and doubtless some of them would object to the use of the word "language" in such a connection. In all matters of real natural history, however (real, that is, in the sense of dealing at first hand with individual birds or beasts), I am much more inclined to listen to old Tomah, who says, when I ask him whether animals can talk: "Talk? Course he kin talk! Eve'ting talk in hees own way. Hear me now make-um dat young owl talk." And, stepping outside the circle of camp-fire light, Tomah utters a hoot, which is answered at a distance every time he tries it. After parleying with the stranger in this tentative fashion, Tomah sends forth a different call; and immediately, as if in ready acceptance of an invitation, a barred owl glides like a gray shadow into a tree over our heads. I have heard that same old Indian use horned-owl talk, wolf and beaver and woodpecker talk, and several other dialects of the wood folk, in the same fascinating and convincing way.

One must judge, therefore, that most cries of the day or night have their meaning, if only one knows how to hear them; yet they constitute but a part, and probably a very small part, of the animal's habitual communication with his fellows. The bulk of it appears to be of that silent kind which passed between Don and Nip, and which, I have reason to believe, is the common language of the whole animal kingdom.

To prove such a matter is plainly impossible. Even to investigate it frankly is to enter a shadowy realm between the conscious and subconscious states, where no process can be precisely followed, and where the liability to error is always present. Let us therefore begin on familiar ground by examining certain phenomena which we cannot explain, to be sure, but

which have been observed frequently enough to give us confidence that we are dealing with realities. I refer especially to that curious warning or "feeling" of impending danger, which is supposed (erroneously, I think) to depend upon the so-called sixth sense of animals and men.

Chumfo, the Supersense

III

FOR the word *chumfo* I am indebted to a tribe of savages living near Lake Mweru, in Africa, and am grateful to them not only for naming a thing which has no name in any civilized language, but also for an explanation of its function in the animal economy. We shall come to the definition of the word presently, after we have some clear notion of the thing for which the word stands. As Thomas à Kempis says, if I remember correctly, "It is better to feel compassion than to know how to define it."

By way of approach to our subject, let it be understood that *chumfo* refers in a general way to the animal's extraordinary powers of sense perception, which I would call his "sensibility" had not our novelists bedeviled that good word by making it the symbol of a false or artificial emotionalism. Every wild creature is finely "sensible" in the true meaning of the word, his sensitiveness being due to the fact that there is nothing dead or even asleep in nature; the natural animal or the natural man is from head to foot wholly alive and awake. And this because every atom of him, or every cell, as a biologist might insist, is of itself sentient and has the faculty of perception. Not till you understand that first principle of *chumfo* will your natural history be more than a dry husk, a thing of books or museums or stuffed skins or Latin names, from which all living interest has departed.

I am sometimes asked, "What is the most interesting thing you find in the woods?" the question calling, no doubt, for the name of some bird or beast or animal habit that may challenge our ignorance or stir our wonder. The answer is, that whether you search the wood or the city or the universe, the only interesting thing you will ever find anywhere is the thrill and mystery of awakening life. That the animal is *alive,* and alive in a way you ought to be but are not, is the last and most fascinating discovery you are likely to make in nature's kingdom. After years of intimate observation, I can hardly meet a wild bird or beast even now without renewed wonder at his aliveness, his instant response to every delicate impression, as if each moment brought a new message from earth or heaven and he must not miss it or the consequent enjoyment of his own sensations. The very sleep of an animal, when he seems ever on the thin edge of waking, when he is still so in touch with his changing world that the slightest strange sound or smell or vibration brings him to his feet with every sense alert and every muscle ready,—all this is an occasion of marvel to dull men, who must be called twice to breakfast, or who meet the violent clamor of an alarm-clock with the drowsy refrain:

> Yet a little sleep, a little slumber,
> A little more folding of the hands to sleep!

You will better understand what I mean by the animal's aliveness, his uncloying pleasure in the sensation of living, if you can forget any tragical theories or prejudices of animal life which you have chanced to read, and then frankly observe the first untrammeled creature you meet in the outdoor world. Here at your back door, for example, is a flock of birds that come trooping from the snowy woods to your winter feast of crumbs. See how they dart hither and yon between mouthfuls, as if living creatures could not be still or content with any one thing, even a good thing, in a world of endless variety. Look again, more closely, and see how they merely taste of the abundance on your table, and straightway leave it for a morsel that the wind blows from under their beaks, and that they are bound to have if it takes all winter. Every other minute they flit to a branch above the table, look about alertly, measure the world once more, make sure of the dog that he is asleep, and of the sky that it holds no hawk; then they wipe their bills carefully, using a twig for a napkin, and down to the table they go to begin all over again. So every bite is for them a feast renewed, a feast with all the spices of the new, the fresh, the unexpected and the adventurous in it.

Or again, when you enter the wilderness remote from men, here is a deer slipping shadow-like through the shadowy twilight, daintily tasting twenty varieties of food in as many minutes, and keeping tabs on every living or moving or growing thing while she eats; or a fox, which seems to float along like thistle-down in the wind, halting, listening, testing the air-smells as one would appreciate a varied landscape, playing Columbus to every nook or brush-pile and finding in it something that no explorer ever found before. Such is the natural way of a fox, which makes a devious trail because so many different odors attract him here or there.

In fine, to watch any free wild creature is to understand the singing lines from "Saul":

> How good is man's life, the mere living, how fit to employ
> All the heart and the soul and the senses forever in joy!

It is to understand also the spirit of Browning, who is hardly a world poet, to be sure, but who has the distinction of being the only famous poet who is always alive and awake. Homer nods; Dante despairs and mourns; Shakespeare has a long period of gloom when he can write only terrible tragedies of human failure; other great poets have their weary days or melancholy hours, but Browning sings ever a song of abounding life. Even his last, the Epilogue to "Asolando," is not a swan-song, like Tennyson's; it is rather a bugle-call, and it sounds not the "taps" of earth, but the "reveille" of immortality. But we are wandering from our woodsy trail.

Those who make an ornithology of mere feathers, or who imagine they know an animal because they know what the scientists have said about him, see in this instant responsiveness of the wild creature only a manifestation of fear, and almost every book of birds or beasts repeats the story of terror and tragedy. Yet every writer of such books probably owns a dog that displays in less degree (because he is less alive) every single symptom of the wild creature, including his alleged fears; and the dog, far from leading a tragic or terror-governed life, is hilariously disposed to make an adventure or a picnic of every new excursion afield. Moreover, if one of these portrayers of animal fears or tragedies has ever had an adventure of his own; if he has penetrated a wild region on tiptoe, or run the white rapids in a canoe, or heard the wind sing in his ears on a breakneck gallop across country, or trailed a bear to his covert, or hunted bandits in the open, or followed the bugles when they blew for war,—then he must know well that these unforgetable moments, when a man's senses all awaken and his nerves tingle and he treads the earth like a buck in spring, are the only times in a man's dull life when he feels himself wholly alive and a man. That a naturalist should forget this when he sees an alert wild animal, and deny his dog and his own experience of life by confounding alertness with fear, is probably due not so much to his own blindness as to his borrowed notions, such as the "struggle for existence," the "reign of terror," and other hallucinations which have been packed into his head in the name of science or natural history.

Just to take a walk with your dog may be a revelation to you, as most simple matters are when you dare to view them for yourself without prejudice. In fact, what is any revelation or discovery but seeing things as they are? This daily walk over familiar ground, which bores you because you must take it for exercise, leaving out of it the fun which is the essential element of any exercise, is to your dog another joyous and expectant exploration of the Indes. He explored the same ground yesterday, to be sure, but all sorts of revolutions and migrations take place overnight; and if he be a real dog, not a spoiled pet, he will uncover more surprises on the familiar road than you uncover in the morning newspaper. Follow him sympathetically as he finds endlessly interesting things where you find little but boredom, and you may learn that there are three marked differences between you and a normal animal: first, that he keeps the spirit of play, which you have well-nigh lost; second, that he lives in his sensations and is happy, while you dwell mostly in your thoughts and postpone happiness for the future; and third, that he is alive now, every moment, and you used to be alive the day before yesterday.

In all this you are simply typical of our unnatural civilization, which begins to enumerate or describe us as so many thousand "souls," doubtless because every single body of us is moribund or asleep. Consider our noses, for example. They are the seat of a wonderful faculty, more dependable than sight or hearing; they are capable of giving us sensations more varied than those of color, and almost as enjoyable as those of harmony; they can be easily trained so as to recognize every tree and plant and growing thing by its delicate fragrance; they would add greatly to our safety and convenience, as well as to our enjoyment, did we use them as nature intended; yet so thoroughly neglected are they, as a rule, that it takes a burning rag or a jet of escaping coal-gas to rouse them to the immense and varied world of odors in which an animal lives continually.

At present I am the alleged owner of a young setter, Rab, who has reached the stage of development when he thinks he owns me. For after you have properly trained a dog, there comes a brief time when he discovers with joy that he can make you do things for him; and then he is like a child who discovers that he can make you talk (that is, show some sign of life) by asking you questions. And this young setter has, I am convinced, a very low opinion of human aliveness, since there is never a day when he does not give me a hint that he considers me a poor cripple— excepting, perhaps, the rare days when I take him woodcock-hunting, when he mildly approves of me.

One night, after waiting a long time at my feet till I should become animate, Rab followed me hopefully into the dark kitchen, where he had never before been allowed to go. I heard his steps behind me only as far as the door, and thought he was afraid to enter; but when I turned on the light, there he stood in the doorway, "frozen" into a beautiful point, his head upturned to a bell and battery on the wall. Presently a mouse hopped from the battery into a waste-basket on the floor, and Rab pointed the thing stanchly, his whole body quivering with delight when a faint odor stole to his nostrils or a rustle of paper to his ears. From the basket the mouse streaked to the coalhod, and Rab pointed that, too, and then a crack under a door, and a yawning closet drawer. "My bones!" he said, trying to claw the drawer open. "This is a good place; something is alive here!" Then he came over and sat in front of me, looking up in my face, his head twisted sideways, to demand why I had so long kept him out of the only part of the house that was not wholly dormant.

Though the adventure was twelve months ago, "age cannot wither it, nor custom stale its infinite variety," for every night Rab comes to me where I am reading, nudges my elbow, paws my book, sits in front of me to yawn hugely, gives me no peace, in short, till I follow him to the kitchen for another go at the waste-basket, the coal-hod and all the closet drawers. It is of no avail to leave the doors open; he will not go unless I join him in the enjoyment of something that lives. A short time since we flushed another mouse in the cellar, whither he followed me to look after the furnace; and now he pesters me by day as well as by night whenever he finds me, as he thinks, wrapped in unmanly lethargy.

With all his vivacity this young setter, who seems so alert among mere men, becomes a dull creature the moment you compare him with his wild kindred. He cannot begin to interpret the world through his senses as they instantly interpret it, and he would starve to death where they live on the fat of the land. Once on a forest trail I came upon him pointing stanchly at the edge of a little opening that lumbermen had made for yarding logs. Just across the opening, where a jumper road joined the yard, stood a noble buck, and he was "pointing," too, for he was face to face with such a creature as he had never before seen. Both animals were like statues, so motionless did they stand; but there was this difference, that the dog rested solidly on earth and might have been carved from marble, while the buck seemed to rest on air and to be compounded of some ethereal essence. His eyes fairly radiated light and color. The velvet on his antlers seemed to grow as I looked upon it, like the velvet moss in which the fairies are said to rest. Every hair of him from nose to tail tip was gloriously alive. A moment only he stood, but long enough for me to carry a picture of him forever afterward; then he bounded up the old road, and Rab came running over to ask me what new thing he had discovered.

This marvelous alertness of the natural animal is commonly attributed to the fact that his physical senses are more acute than ours; but that is true only of some particular sense of a certain creature. The wolf's nose, the deer's ear, the vulture's eye,—these are probably keener than any similar human organ; but, on the other hand, a man's eye is very much keener for details than the eye of wolf or deer; his senses of touch and taste are finer than anything to be found among the lower orders, and the average of his five senses is probably the highest upon earth. Yet the animal is more responsive to impressions of the external world, and this is due, I think, to the fact that he lives more in his sensations; that he is not cumbered, as we are, by inner phenomena; that he is free from pain, care, fear, regret, anxiety and other mental complexities; that he is accustomed to hear (that is, to receive vibrations) through his feet or his skin quite as much as through his ears; and that his whole body, relaxed and at ease, often becomes a first-class receiving instrument for what we call sense impressions.

As an example of this last, note the wild creature's response to every change of air pressure, a response so immediate and certain that one might accurately forestall the barometer by observing the action of birds, or even of chickens, which anticipate a storm long before your face has noted the moist wind or your eye the rain-cloud. In the winter woods I have often seen a deer feeding greedily (quite at variance with his usual dainty tasting) at an hour long before or long after his accustomed time, and as I traveled wider I would find other deer doing the same thing. I used to wonder at this, till I noticed that such unusual action was always followed by a storm—not an ordinary brief snowfall, to which deer pay little attention beyond seeking shelter while it lasts, but a severe storm or blizzard, during

which most animals lie quiet for a whole day, or even two or
three days, without stirring abroad for food.

Whether an element of forethought enters into this act of
"stuffing" themselves before a storm, or whether it is wholly
instinctive, like the bear's change of diet before he dens for his
winter sleep, is a question which does not concern us, since
nobody can answer it. In either case the deer have felt, as
surely as our most sensitive instruments, not only the decreasing
pressure of the air but also its increasing moisture.

That such sensitiveness is not of any one organ, but rather of
the whole body, becomes more evident when we study the
lower orders—fishes, for example, which may winter under a
dozen fathoms of water and a two-foot blanket of ice, but which
nevertheless respond to the changing air currents far above
their heads. Once on a northern lake, in March, I kept tabs on
some trout for fourteen consecutive days, and it seemed that
they moved from deep to shallow water or back again whenever
the wind veered to the proper quarter. I had a water-hole cut in
the thick ice, and, finding a trout under it one day, I kept a
couple of lines with minnows there constantly. The hole was in
a shallow place, over a sandy bottom, and by putting my face to
the opening, with a blanket over my head to exclude the upper
light, I could dimly see the shadows move in from deep water.
On six scattering days, when the wind came light or strong from
the south, two of the days bringing snow, the trout evidently
moved shoreward, since I caught them abundantly, as many as
I needed and some to spare; but on the other eight days, some
clear and some stormy, when the wind was north of east or west,
not a trout was seen or caught, and only once was a minnow of
mine missing.

It may be said that the trout simply followed their food-
supply; but I doubt this, since trout apparently feed very little in
winter; and in formulating any theory of the matter one must
account for the fact that big fish or little fish moved shoreward
whenever the wind blew south. The phenomenon may appear
less foreign to our experience, though not less mysterious to our
reason, if we remember that an old wound or a corn may by its
aching foretell a storm, or that a person suffering from nervous

prostration may by his sudden depression know that the barometer will soon be falling.

The same bodily sensitiveness appears unchanged in our domestic animals. I once saw a deer and her two fawns kneel down in the woods, and watched them in astonishment as they rested for some time on their knees, as if in supplication; then the ground rocked under me, and I knew that their feet had felt the tremor of an earthquake long before I was sensible of it. Such an observation seemed wonderful to me till I learned that our sheep are equally sensitive in their elastic hoofs, and that our pigs respond not only to vibration of earth or air, but also to some finer vibration set in motion, apparently, by human excitement.

Moreover, I have known one dog, old and half deaf, that, whether asleep or awake, would respond to the faint tremor of his master's automobile before it came into sight or hearing. And what there is in the tremor of one machine to distinguish it from another of the same make and power is something that the unaided ear can hardly measure. The dog lived under a hilltop, on a highway over which scores of automobiles passed daily; and on holidays, when his master was at home, the scores would increase to hundreds. He would sleep for hours on the veranda, paying no heed to the noise or smell or dust of the outrageous things, till suddenly he would jump up, bark, and start for the gate; and in a moment or two we would see the master's auto rise into sight over the brow of the hill.

The instant response of deer or dog to minute external impressions, though startling enough, is probably wholly physical, a matter of vibrations on one side and of nerves on the other; but there are other phenomena of sensitiveness (and these bring us nearer to our trail of animal communication) for which it is much harder to find a satisfactory physical explanation. Such is the feeling or warning of unsensed danger, or the premonition that some one unseen and unheard is approaching—a phenomenon which seems to be common among animals, to judge from repeated observation, and which appears often enough in human beings to make not only the inquisitive

Society for Psychical Research but almost every thoughtful man or woman take some note of it.

For example, a man awake in his bed sees his son, whom he thinks safely ashore in a foreign country, fall overboard from a steamer to his death, at the very hour when the son did fall overboard, as was afterward learned. Or a woman, the wife of a sea-captain, sitting on the veranda at home in the bright moonlight, sees the familiar earth vanish in a world of water, and looks suddenly upon her husband's ship as it reels to the gale, turns over to the very edge of destruction, and then rights itself with half its crew swept overboard—and all this while the precise event befell a thousand miles away. Such things, which spell a different kind of sensitiveness from that with which we are familiar, have happened to people well known to me; but as they have happened to others also, and as almost every town or village has a convincing example of its own, I forbear details and accept the fact, and try here to view or understand it as a *natural* phenomenon.

At first you may strongly object even to my premise, calling it incredible that sense-bound mortals should feel a danger that their eyes cannot see or their ears hear; but there are at least two reasonable answers to your objection. In the first place, we are sense-bound only in the sense of limiting ourselves unnecessarily, confining our perception to five habitual modes, shamefully neglecting to cultivate even these, and ignoring the use or the existence of other and perhaps finer means of contact with the external world. Again, it is not a whit more incredible that sensitive creatures, whether brute or human, should feel the coming or going of a person than that they should feel his look or glance, as they certainly do.

This last is no cloudy theory; it is a plain fact which endures the test of observation. Almost any man of strong personality can disturb or awaken a sleeping wild animal simply by looking at him intently; and the nearer the man is the more certain the effect of his gaze on the sleeping brute. The same is true in less degree of most Indians and woodsmen, and of many sensitive women and children, as you may prove for yourself. Go into a room where a sensitive or "high-strung" person is taking a nap—

not sleeping heavily, as most men sleep, but lightly, naturally, as all wild animals take their rest. Make no noise, but stand or sit quietly where you can look intently into the sleeper's face; and commonly, by a change of position or a turning away of the head or a startled opening of the eyes, the sleeper will show that he feels your look and is trying subconsciously to avoid it. The awakening, whether of animals or of men, does not always follow our look, most fortunately, but it happens frequently enough under varying conditions to put the explanation of chance entirely out of the question.

When I was a child I used to sit long hours in the woods alone, partly for love of the breathing solitude, and partly for getting acquainted with wild birds or beasts, which showed no fear of me when they found me quiet. At such times I often found within myself an impression which I expressed in the words, "Something is watching you." Again and again, when nothing stirred in my sight, that curious warning would come; and almost invariably, on looking around, I would find some bird or fox or squirrel which had probably caught a slight motion of my head and had halted his roaming to creep near and watch me inquisitively. As I grew older the "feel" of living things grew dimmer; yet many times in later years, when I have been in the wilderness alone, I have experienced the same impression of being watched or followed, and so often has it proved a true warning that I still trust it and act upon it, even when my eyes see nothing unusual and my ears hear nothing but their own ringing in the silence.

I remember once, when I was sitting on the shore of a lake at twilight, that I began to have an increasing impression that some living thing unseen was near me. At first I neglected it, for I had my eyes on a deer that interested me greatly; but the feeling grew stronger till I obeyed it and rose to my feet. At the first motion came a startling *woof!* and from some bushes close behind me a bear jumped away for the woods. No doubt he had been there some time, watching me or creeping nearer, knowing that I was alive, but completely puzzled by my shapelessness and lack of action. A similar thing has happened several times, in other places and with other animals, and always at a moment when I was most in harmony with the environment and a sharer of its deep tranquillity.[1]

As a child this faculty (if such it be) was as natural as anything else in life; for in childhood we take the world as we find it in personal experience, and nothing is especially wonderful where all is wonder. I then thought no more of feeling the presence of an animal than of hearing him walk when I could not see him; but as I grew older the experience seemed a little odd or "queer," and I never spoke of it to any one till I discovered, first, that the faculty seems to be common among animals, and second, that some Indians (not all) have it and regard it as the most natural thing in the world. Here is how the latter discovery came about:

Simmo and I were calling moose from a lake one moonlit night, with a silent canoe under us, dark evergreen woods at our back, and a little ghost of a beaver-meadow, vague, misty, shadow-filled, immediately before our eyes yet seeming as remote as any drifting cloud. To our repeated call no answer was returned; then we allowed the canoe to drift ashore where it would, and sat listening to the vast silence. I was brought back from my absorption in the fragrance, the harmony, the infinite stillness of the night by feeling the canoe shake and hearing Simmo whisper, "Somet'ing near. Look out!"

Now in a silence like that, the tense, living silence of the wilderness at night, one's ears are as full of tricks as a *puckwudgie*, who is one of the mischievous fairy-folk of the Indians. The whine of a mosquito sounds across the whole lake; or the mutter of a frog becomes like the roar of a bull of Bashan; or suddenly you begin to hear music as the faint vibration of some dry stub, purring in the unfelt air currents, turns to the booming of a mighty church organ. At such a time, unless one has trained his senses to ignore the obvious (the Indian secret of seeing and hearing things), one soon becomes confused, uncertain of the borderland between the real and the imaginary; so presently I turned to Simmo and whispered, "You hear him?"

The Indian shook his head. "No hear-um; just feel-um," he

---

[1] For further example and analysis of the matter, see pp. 120–122.

said; and again we settled down to watch. For several minutes we questioned the woods, the lake, the meadow; but nothing stirred, not a sound broke the painful quiet, the while we both felt strongly that some living thing was near us. Then a shadow moved from the darker shadow of an upturned root, only a few yards away, and a great bull stood alert on the open shore.

I thought then, and I still think, that besides our ordinary five senses we have a finer faculty which I must call, for lack of a better term, the sense of presence; and I explain it on the assumption that every life recognizes and attracts every other life by some occult force, as dead matter (if there be such a thing) attracts all other matter by the mysterious force of gravitation.

Doubtless I have been many times watched in the woods when I did *not* know or feel it; and certainly I have often had my eyes upon an unconscious bird or beast when, though I fancied he grew uneasy under my scrutiny, he did not have enough sense of danger to run away—possibly because no real danger threatened him, the eyes that looked upon him being friendly or merely curious.

Once on a hardwood ridge I came upon a buck lying asleep in open timber, and stood with my back against a great sugar-maple, observing him for five or six minutes before he stirred. No, I was not trying to awaken him by a look, or to stage any other experiment. I was simply enjoying a rare sight, noting with immense interest that this wild creature, whom I had always seen so splendidly alert, could nod and blink like any ordinary mortal. So near was he that I could mark the drowsiness of his eyes, the position of his feet, the swelling of his sides as he breathed. I happened to be counting his respirations in friendly fashion, comparing them with my own, when suddenly his head turned to me; his eyes snapped wide open, and they were looking straight into mine. Apparently his feeling or subconscious perception had warned him where danger was; but still his eyes could not recognize it, standing there in plain sight. For to stand motionless without concealment is often the best way to deceive a wild animal, which habitually associates life with motion.

No more drowsiness for that buck! He was startled, plainly

enough; but he rose to his feet very stealthily, not stirring a leaf, and stood at tense attention. When he turned his head to look over his shoulder behind him I raised my field-glass, for I wanted to read his thought, if possible, in his eyes. As he turned his gaze my way again his nose seemed to sweep my face. It rested there a moment full on the lens of my glass, moved on, and returned for a longer inspection. Then he glided past, still without recognizing me, testing the air at every springy step, harking this way, looking that way, and disappearing at last as if he trod on eggs.

From such experiences I judge that the feeling of unsensed danger, or the more subtle feel of a living thing, is as variable in the animals as are their instincts or their social habits. It may be dull in one creature and keen in another of the same species, or alternately awake and asleep in the same creature; but there is no longer any doubt in my mind that it is a widespread gift among birds and beasts. When it appears occasionally among men, therefore, it is to be regarded not as uncanny or queer, but, like the sure sense of direction which a few men possess, as a precious and perfectly natural inheritance from our primitive ancestors. That it skips a dozen generations to alight on an odd man here or there, like a storm-driven bird on a ship at sea, is precisely what we might expect of heredity, which follows a course that seems to us erratic or at times marvelous, as geometry appears to an Eskimo, because we do not yet understand its law or working principle.

Among savage tribes, who live a natural outdoor life in close contact with nature, the perception of danger or of persons beyond the ordinary sense range is much more common than among civilized folk. Almost every explorer and missionary who has spent much time with African natives, for example, has noticed that the Blacks have some mysterious means of knowing when a stranger is approaching one of their villages—mysterious, that is, because it does not depend on runners or messengers or any other of our habitual means of communication. A recent observer of these people has at last offered an explanation of the matter, with many other impressions of native philosophy, in a work to which he gives the suggestive title of

Thinking Black. Of the scores of books on Africa which I have read, this is the only one to show more than a very superficial knowledge of the natives, and the reason is apparent. The author thinks, most reasonably, that you cannot possibly know the native or understand his customs until you know his thought, and that the only trail to his thought is through the language in which the thought is expressed. Therefore did he study the speech as a means of knowing the man; and herein he is in refreshing contrast to other African travelers and hunters I have read, who spend a few months or weeks in a white man's camp, knowing the natives about as intimately as lovers know the moon, and then babble of native customs or beliefs or "superstitions,"—as if any rite or habit could be understood without first understanding the philosophy of life from which it sprang, as a flower from a hidden seed.

This rare observer, who knows how the native thinks because he perfectly understands the native language, tells us[2] that the Blacks clearly recognize the power of animals and of normal men to know many things beyond their sense range; that they give it a definite name, and explain it in a way which indicates an astonishing degree of abstract thought on the part of those whom we ignorantly call unthinking savages. According to these natives, every natural animal, man included, has the physical gifts of touch, sight, hearing, taste, smell, and *chumfo*. I must still use the native term because it cannot be translated, because it implies all that we mean by instinct, intuitive or absolute knowledge and (a thing which no other psychology has even hinted at) the process by which such knowledge is acquired.

This *chumfo* is not a sixth or extra sense, as we assume, but rather the unity or perfect co-ordination of the five senses at their highest point. I may illustrate the matter this way, still following Crawford, whose record contains many curious bits of observation, savage philosophy, woods lore and animal lore, many of them written by the camp-fire and all jumbled together pell-mell:

For ordinary perception at near distances the eye or the ear

---

[2] Crawford, *Thinking Black* (1904).

is sufficient, and while engaged in any near or obvious matter the five senses work independently, each busy with its own function. But when such observation is ended or at fault, and the man retreats, as it were, into his inner self, then in the quiet all the senses merge and harmonize into a single perfect instrument of perception. (Here, in native dress, is nothing more or less than the psychologist's subconscious self, with its mysterious working.) At such moments the whole animal or the whole man, not his brain and senses alone, becomes sensitive to the most delicate impressions, to inaudible sounds or vibrations, to unseen colors, to unsmelled odors or intangible qualities,—to a multitude of subtle messages from the external world, which are ordinarily unnoticed because the senses are ordinarily separate, each occupied with its particular message. So when a sleeping animal is suddenly aware that he must be alert, he does not learn of approaching danger through his ears or nose, but through *chumfo,* through the perfect co-ordination of all his senses working together as one. In the same way a wandering black man always knows where his hut or camp is; he holds his course on the darkest night, finds his way through a vast jungle, goes back to any spot in it where he left something, and often astonishes African travelers by getting wind of their doings while they are yet far distant.

Such is the native philosophy; and the striking feature of it is, that it is not superstitious but keenly observant, not ignorant but rational and scientific, since it seems to anticipate our latest biological discoveries, or rather, as we shall see, a philosophy which rests upon biological science as a foundation.

Perhaps the first suggestion that the native may have reason in his theory comes from the extraordinary sensitiveness of certain blind people, who walk confidently about a cluttered room, or who sort the family linen after it returns from the laundry. These blind people say, and think, that they avoid objects by feeling the increasing air pressure as they approach, or that they sort the family linen by smell; but it appears more likely that a greater unity and refinement of all their senses results from their living in darkness.

That our human senses have unused possibilities, or that we

may possibly possess extra senses of which we are not conscious, may appear if you study the phenomenon of hearing, especially if you study it when you hear a strange sound in the woods or in the house at night. It is assumed that we always locate a sound by the ear, and that we determine its volume or distance by our judgment from previous experiences; but that, I think, is a secondary and not a primary process. When we act most naturally we seem to locate a sound not by search or experiment, but instantly, instinctively, absolutely; and then by our ear or our judgment we strive to verify our first *chumfo* impression.

As a specific instance, you are lying half asleep at night when a faint, strange sound breaks in upon your consciousness. If you act naturally now, you will nine times out of ten locate the sound on the instant, without bringing your "good ear" to bear upon it; but if you neglect or lose your first impression, you may hunt for an hour and go back to bed without finding the source of the disturbance. Or you are traveling along a lonely lane at night, more alive than you commonly are, when a sudden cry breaks out of the darkness. It lasts but a fraction of a second and is gone; yet in that fleeting instant you have learned three things: you know the direction of the cry; you know, though you never heard it before, whether it is a loud cry from a distance or a faint cry from near at hand; and you are so sure of its exact location that you go to a certain spot, whether near or far, and say, "That cry sounded here."

So much if you act naturally; but if you depend on your ears or judgment, as men are apt to do, then you are in for a long chase before you locate the cry of a bird or a beast or a lost child in the night.

It is easy to make mistakes here, for we are so cumbered by artificial habits that it is difficult to follow any purely natural process; but our trail becomes clearer when we study the matter of hearing among the brutes. Thus, your dog is lying asleep by the fire when a faint noise or footstep sounds outside. Sometimes, indeed, there is no audible sound at all when he springs to his feet; but no sooner is the door opened for him than he is around the house and away, heading as straight for the disturbance as if he knew, as he probably does, exactly where to find it. Yet your dog is, I repeat, a very dull creature in comparison with his wild kindred. Their ability to locate a sound is almost unbelievable, not because they have more delicate ears (for the human ear is much finer, being sensitive to a thousand inflections, tones, harmonies, which are meaningless to the brute), but because of what the Blacks call their better *chumfo* or what we thoughtlessly call their stronger instincts.

This has been strongly impressed upon me at times when I have tried to call a moose in the wilderness. If you seek these animals far back where they are never hunted—a difficult matter nowadays—the bulls answer readily enough, or sometimes too readily, as when one big brute chased me to my canoe and gave me a hatless run for it; but in a much-hunted region they are very shy and come warily to a call. The best way to see them in such a place is to call a few times at night, or until you get an answer, and then go quickly away before the bull comes near enough to begin circling suspiciously. At daybreak you are very apt to find him waiting; and the astonishing thing is that he is waiting at the very spot where you used your trumpet.

The place you select for calling may be a tiny bog in a vast forest, or a little, nameless beaver-meadow by a lake or river. It is like many other such places, near or far, and the bull may come from a distance, crossing lakes, rivers, bogs and dense forest on his way; but he never seems to make a mistake or to be at a loss in locating the call. On a still night I have heard a bull answer me from a mountain five or six miles away; yet in the morning there he was, waiting expectantly for his mate near the bit of open shore where I had called him; and to reach that spot he must either have crossed the lake by swimming, a distance of two miles, or else have circled it on a wide détour. That he should come such a distance through woods and waters, and pick the right spot from a hundred others on either side, seems to me not a matter of ears or experience but of *chumfo*, or absolute knowledge.

Another and more interesting verification of the *chumfo* philosophy is open to any man who will go quietly through the big woods by moonlight, putting himself back amid primal or animal conditions, and observing himself closely as he does so. The man who has not traveled the wilderness alone at night has a vivid and illuminating experience awaiting him. He is amazed, so soon as he overcomes the first unnatural feeling of fear, to find how alive he is, and h'ow much better he can hear and smell than ever he dreamed. At such a time one's whole body seems to become a delicately poised instrument for receiving sense impressions, and one's skin especially begins to tingle and creep as it wakes from its long sleep. Nor is this "creeping" of the skin strange or queer, as we assume, but perfectly natural. The sensations which we now ignorantly associate with fear of the dark (a late and purely human development; the animal knows it not) are in reality the sensations of awakening life.

Possibly we may explain this supersensitiveness of the skin, when life awakens in it once more and it becomes for us another and finer instrument of perception, by the simple biological fact that every cell of the multitudes which make up the human body has a more or less complete organization within itself. Moreover, as late experiments have shown, a cell or a tissue of cells will live and prosper in a suitable environment when completely separated from the body of which it was once a part. These human cells inherit certain characteristics common to all animal cells since life began; and it is not improbable that they inherit also something of the primal cell's sensibility, or capacity to receive impressions from the external world. This universal cell-function was largely given over when the animal (a collection of cells) began to develop special organs of touch, sight, and hearing; but there is no indication that the original power of sensibility has ever been wholly destroyed in any cell. It is, therefore, still within the range of biological possibility that a man should hear with his fingers or smell with his toes, since every cell of both finger and toe once did a work corresponding to the present functions of the five animal senses.

Before you dismiss this as an idle or impossible theory, try a simple experiment, which may open your eyes to the reality of living things. Go to a greenhouse and select a spot of bare earth under a growing rose-bush. Examine the surface carefully, then brush and examine it again, to be sure that not a root of any kind is present. Now place a handful of good plant-food on the selected spot, and go away to your own affairs. Return in a week or so, brush aside your "bait," and there before your eyes is a mass of white feeding-roots where no root was before. In some way, deep under the soil, the hungry cells have heard or smelled or felt a rumor of food, and have headed for it as surely as a dog follows his nose to his dinner.

Do plants, then, *know* what they do when they turn to the light, and is there something like consciousness in a tree or a blade of grass? That is too much to assert, though one may think or believe so. No man can answer the question which occurs so constantly to one who lives among growing things; but you can hardly leave your simple experiment without formulating a theory that even the hidden rootlets of a rose-bush have something fundamentally akin to our highly developed sensibility. At present some biologists are beginning to assert, and confidently, though it is but an opinion, that there is no dead matter in the world; that the ultimate particles of which matter is composed are all intensely alive. And if alive, they must be sentient; that is, each must have an infinitesimal degree of feeling or sense perception.

To return from our speculation, and to illustrate the *chumfo* faculty from human and animal experience: I was once sitting idly on a Nantucket wharf, alternately watching some hermit-crabs scurrying about in their erratic fashion under the tide, and an old dog that lay soaking himself in the warm sunshine. Just behind us, the only inharmonious creatures in the peaceful scene, some laborers were unloading rocks from a barge by the aid of a derrick. For more than an hour, or ever since I came to the wharf, the dog lay in the same spot, and in all that time I did not see him move a muscle. He was apparently sound asleep. Suddenly he heaved up on his rheumatic legs, sniffed the air alertly, and turned his head this way, that way, as if wary of something.

The human labor had proceeded lazily, for the day was warm; there was no change in the environment, so far as I could discern; the only sounds in the air were the sleepy lap of wavelets and the creaking of pulleys; yet my instant thought was, "That dog is frightened; but at what?" After a few moments of watching he moved off a dozen yards and threw himself flat on his side to sleep again. His body was hardly relaxed when a guy-rope parted, and the iron-bound mast of the derrick crashed down on the wharf.

It was certainly "touch and go" for me; I felt the wind of the thing as it fell, and was almost knocked off the wharf; but I was not thinking then of my own close call. With my interest at high pitch I examined the mast, and found it lying squarely athwart the impression left by the dog in the dust of the road.

"Merely a coincidence," you say; which indicates that we are apt to think alike and in set formulas. That is precisely what I said at the time—a mere coincidence, but a startling one, which made me think of luck (a most foolish notion) and wonder why luck should elect to light on a worthless old dog and take no heed of what seemed to me then a precious young man. But I have since changed my mind; and here is one of the many observations which made me change it.

Years afterward my Indian guide, Simmo, was camped with a white man beside a salmon river. It was a rough night, and a storm was roaring over the big woods. For shelter they had built a bark *commoosie,* and for comfort a fire of birch logs. At about nine o'clock they turned in, each wrapped in his blanket, and slept soundly but lightly, as woodsmen do, after a long day on the trail. Some time later—hours, probably, for the fire was low, the storm hushed, the world intensely still—the white man was awakened by a touch, and opened his eyes to find his companion in a tense, listening attitude.

"Bes' get out of here quick, 'fore somet'ing come," said the Indian, and threw off his blanket.

"But why—what—how do you know?" queried the white man, startled but doubting, for he had listened and heard nothing.

The Indian, angered as an Indian of the woods always is when you question or challenge his craft, made an impatient gesture. "Don' know how; don' know why; just know. Come!" he called sharply, and the white man followed him away from the camp toward the river, where it was lighter. For several minutes they stood there, like two alert animals, searching the dark woods with all their senses; but nothing moved. The white man was beginning another fool question when there came a sudden dull crack, a booming of air, as a huge yellow-birch stub toppled over the fire and flattened the *commoosie* like a bubble.

"Dere! Das de feller mus' be comin'," said Simmo. "By cosh, now, nex' time Injun tell you one t'ing, p'r'aps you believe-um!" And, as if it were the most natural thing in the world, he stepped over death-and-destruction, kicked aside some rubbish, and lay down to sleep where he was before.

"How do I explain it?" I don't. I simply recognize a fact which I cannot explain, and which I will not blink by calling it another coincidence. For the fact is, as I judge, that a few men and many animals exercise some extra faculty which I do not or cannot exercise, or have access to some source of information which is closed to me. When I question the gifted men or women who possess this faculty, or what's-its-name, I find that they are as much in the dark about it as I am. They know certain things without knowing how they learn; and the only word of explanation they offer is that they "feel" thus and so—perhaps as a horse feels when he is holding the right direction through a blinding snowstorm, as he does hold it, steadily, surely, if you are wise enough not to bother him with the reins or your opinions.

Simmo is one of these rare men. At one moment he is a mere child, so guileless, so natural, so innocent of worldly wisdom, that he is forever surprising you. Once when his pipe was lost I saw him fill an imaginary bowl, scratch an imaginary match, and puff away with a look of heavenly content on his weathered face. So you treat him as an unspoiled creature, humoring him, till there is difficulty or danger ahead, or a man's work to be done, when he steps quietly to the front as if he belonged there. Or you may be talking with him by the camp-fire, elaborating some wise theory, when he brushes aside your book knowledge as of no consequence and suddenly becomes a philosopher, proclaiming a new or startling doctrine of life in the sublimely unhampered way of Emerson, who finished off objectors by saying, "I do not argue; I know." But where Emerson gives you a mystical word or a bare assertion which he cannot possibly prove, Simmo has a disconcerting way of establishing a challenged doctrine by a concrete and undeniable fact.

One misty day when we were astray in the wilderness, he and I, we attempted to travel by getting our compass bearings from the topmost twigs of the evergreens, which slant mostly in one direction. After blundering around for a time without getting any nearer camp or familiar landmarks, Simmo remarked: "Dese twigs lie like devil. I guess I bes' find-um way myself." And he did find it, and hold it even after darkness fell, by instinctive feeling. At least, I judge it to have been a matter of feeling rather than of sense or observation, for his only explanation was, "Oh, w'en I goin' right I feel good; but w'en I goin' wrong I oneasy."

This natural feeling or impression of things beyond the range of sight, this extra sense, or *chumfo* unity of all the senses, is probably akin to another feeling by which the animal or man becomes aware of distant persons, or of distant moods or emotions. The sleeping dog's alarm beneath the weakened derrick, or the sleeping Indian's uneasiness near the doomed birch-stub, might be explained on purely physical grounds: some tremor of parting fibers, some warning vibration too faint for eardrums but heavy enough to shake a more delicately poised nerve center, reached the inner beast or the inner man and roused him to impending danger. (There is a deal of babble in this explanation, I admit, and still a mystery at the end of it.) But when a man or a brute receives knowledge not of matter, but of minds or spirits like his own; when a mother knows, for example, the mental state of a son who is far away, and when no material vibrations of any known medium can pass between them,—then all sixth-sense theories, which must rest on the impinging of waves upon nerve centers, no longer satisfy or explain. We are in the more shadowy region of thought transference or impulse transference, and it is in this silent, unexplored region that,

as I now believe, a large part of animal communication goes on continually.

That belief will grow more clear, and perhaps more reasonable, if you follow this unblazed trail a little farther.

Natural Telepathy

IV

T<small>HE</small> way of animal communication now grows dimmer and dimmer, or some readers may even think it "curiouser and curiouser," as Alice of Wonderland said when she found herself lengthening out like a telescope. But there is certainly a trail of some kind ahead, and since we are apt to lose it or to wander apart, let us agree, if we can, upon some familiar fact or experience which may serve as a guiding landmark. Our general course will be as follows: first, to define our subject, or rather, to make its meaning clear by illustration; second, to examine the reasonableness of telepathy from a natural or biological viewpoint; and finally, to go afield with eyes and minds open to see what the birds or the beasts may teach us of this interesting matter.

It seems to be fairly well established that a few men and women of uncommonly fine nervous organization (which means an uncommonly natural or healthy organization) have the power of influencing the mind of another person at a distance; and this rare power goes by the name of thought transference, or telepathy. The so-called crossing of letters, when two widely separated persons sit down at the same hour to write each other on the same subject, is the most familiar but not the most convincing example of the thing. Yes, I know the power and the example are both challenged, since there are scientists who deny telepathy root and branch, as well as scientists who believe in it implicitly; but I also know something more convincing than any second-hand denial or belief, having at different times met three persons who used the "gift" so freely, and for the most part so surely, that to ignore it would be to abandon confidence in my own sense and judgment. I am not trying, therefore, to investigate an opinion, but to understand a fact.

To illustrate the matter by a personal experience: For many years after I first left home my mother would become "uneasy in her mind," as she expressed it, whenever a slight accident or danger or sickness had befallen me. If the event were to me serious or threatening, there was no more doubt or uneasiness on my mother's part. She would know within the hour that I was in trouble of some kind, and would write or telegraph to ask what was the matter.

It is commonly assumed that any such power must be a little weird or uncanny; that it contradicts the wholesome experience of humanity or makes fantastic addition to its natural faculties; and I confess that the general queerness, the lack of balance, the Hottentotish credulity of folk who dabble in occult matters give some human, if not reasonable, grounds for the assumption. Nevertheless, I judge that telepathy is of itself wholly natural; that it is a survival, an age-old inheritance rather than a new invention or discovery; that it might be exercised not by a few astonishing individuals, but by any normal man or woman who should from infancy cultivate certain mental powers which we now habitually neglect. I am led to this conviction because I have found something that very much resembles telepathy in frequent use throughout the entire animal kingdom. It is, as I think and shall try to make clear, a natural gift or faculty of the animal mind, which is largely subconscious, and it is from the animal mind that we inherit it; just as a few woodsmen inherit the animal sense of direction, and cultivate and trust it till they are sure of their way in any wilderness, while the large majority of men, dulled by artificial habit, go promptly astray whenever they venture beyond beaten trails.

That the animals inherit this power of silent communication over great distances is occasionally manifest even among our half-natural domestic creatures. For example, that same old setter of mine, Don, who introduced us to our fascinating subject, was left behind most unwillingly during my terms at school; but he always seemed to know when I was on my way home. For months at a stretch he would stay about the house, obeying my mother perfectly, though she never liked a dog; but on the day I was expected he would leave the premises, paying no heed to orders, and go to a commanding ledge beside the lane, where he could overlook the highroad. Whatever the hour of my coming, whether noon or midnight, there I would find him waiting.

Once when I was homeward bound unexpectedly, having sent no word of my coming, my mother missed Don and called him in vain. Some hours later, when he did not return at his dinner-time or answer her repeated call, she searched for him and found him camped expectantly in the lane. "Oho! wise dog," said she. "I understand now. Your master is coming home." And without a doubt that it would soon be needed, she went and made my room ready.

If the dog had been accustomed to spend his loafing-time in the lane, one might thoughtlessly account for his action by the accident or hit-or-miss theory; but he was never seen to wait there for any length of time except on the days when I was expected. And once (unhappily the last time Don ever came to meet his master) he was observed to take up his watch within a few minutes of the hour when my train left the distant town. Apparently he knew when I headed homeward, but there was nothing in his instinct or experience to tell him how long the journey might be. So he would wait patiently, loyally, knowing I was coming, and my mother would take his dinner out to him.

In many other ways Don gave the impression, if not the evidence, that he was a "mind-reader." He always knew when Saturday came, or a holiday, and possibly he may have associated the holiday notion with my old clothes; but how he knew what luck the day had in store for him, as he often seemed to know the instant I unsnapped his chain in the early morning, was a matter that at first greatly puzzled me. If I appeared in my old clothes and set him free with the resolution that my day must be spent in study or tinkering or farm work, he would bid me good morning and go off soberly to explore the premises, as dogs are wont to do. But when I met him silently with the notion that the day was my day off, to be wasted in shooting or fishing or roving the countryside, then in some way Don caught the notion instantly; he would be tugging at his leash before I reached him, and no sooner was he free than he was all over the yard in mad capers or making lunatic attempts to drag me off on our common holiday before breakfast.

That any dog of mine should obey my word, doing gladly whatever I told him, was to be expected; or that in the field he should watch for a motion of my hand and follow it instantly, whether to charge or hold or come in or cast left or right, was a simple matter of training; but that this particular dog should, unknown to me, enter into my very feeling, was certainly not the result of education, and probably not of sight or sense, as we ordinarily understand the terms. When we were together of an evening before the fire, so long as I was working or pleasantly reading he would lie curled up on his own mat, without ever disturbing me till it was time for him to be put to bed, when he would remind me of the fact by nudging my elbow. But if an hour came when I was in perplexity, or had heard bad news and was brooding over it, hardly would I be away in thought, forgetful of Don's existence on a trail I must follow alone, when his silky head would slide under my hand, and I would find his brown eyes searching my face with something inexpressibly fine and loyal and wistful in their questioning deeps.

Thus repeatedly, unexpectedly, Don seemed to enter into my moods by some subtle, mysterious perception for which I have no name, and no explanation save the obvious one—that a man's will or emotion may fill a room with waves or vibration as real as those streaming from a fire or a lighted candle, and that normal animals have some unused bodily faculty for receiving precisely such messages or vibrations. But we are not yet quite ready for that part of our trail; it will come later, when we can follow it with more understanding.

Should this record seem to you too personal (I am dealing only with first-hand impressions of animal life), here is the story of another dog—not a blue-blooded or highly trained setter, but just an ordinary, doggy, neglected kind of dog—submitted by a scientific friend of mine, who very cautiously offers no explanation, but is content to observe and verify the facts:

This second dog, Watch by name and nature, was accustomed to meet his master much as Don met me in the lane; but he did it much more frequently, and timed the meeting more accurately. He was nearer the natural animal, never having been trained in any way, and perhaps for that reason he retained more of the natural gift or faculty of receiving a message from a distance. His owner, a busy carpenter and builder, had an office in town, and was accustomed to return from his office or work at all hours, sometimes early in the afternoon, and again long after dark. At whatever hour the man turned homeward, Watch seemed to follow his movement as if by sight; he would grow uneasy, would bark to be let out if he happened to be in the house, and would trot off to meet his master about half-way. Though he was occasionally at fault, and sometimes returned to brood over the matter when his master, having started for home, was turned aside by some errand, his mistakes were decidedly exceptional rather than typical. His strange "gift" was a matter of common knowledge in the neighborhood, and occasionally a doubtful man would stage an experiment: the master would agree to mark the hour when he turned homeward, and one or more interested persons would keep tabs on the dog. So my scientific friend repeatedly tested Watch, and observed him to take the road within a few moments of the time when his master left his office or building operations in the town, some three or four miles away.

Thus far the record is clear and straight, but there is one important matter which my friend overlooked, as scientific men commonly do when they deal with nature, their mistake being to regard animals as featureless members of a class or species rather than as individuals. The dog's master always came or went in a wagon drawn by a quiet old horse, and upon inquiry I found that between Watch and the horse was a bond of comradeship, such as often exists between two domestic animals of different species. Thus, the dog often preferred to sleep in the stall near his big chum, or would accompany him to the pasture when he was turned loose, and would always stand by, as if overlooking the operation, when the horse was being harnessed. It may well be, therefore, that it was from the horse rather than from the man that Watch received notice when heads were turned homeward; but of the fact that some kind of telepathic communication passed between two members of the trio there is no reasonable doubt.

Some of my readers may make objection at this point that, though something like telepathic communication appears now and then among the brutes, it should be regarded as merely freakish or sensational, like a two-headed calf; while others will surely ask, "Why, if our dogs possess such a convenient faculty, do they not use it more frequently, more obviously, and so spare themselves manifold discomforts or misunderstandings?"

Such an objection is natural enough, since we judge as we live, mostly by habit; but it has no validity, I think, and for two reasons. First, because such animals as we have thus far seen exercising the faculty (and they are but a few out of many) are apparently normal and sensible beasts, precisely like their less-gifted fellows; and second, because the telepathic power itself, when one examines it without prejudice, appears to be wholly natural, and sane or simple as the power of thought, even of such rudimentary thought as may be exercised in an animal's head. As for emotions, more intense and penetrating than any thought, it is hardly to be questioned that a man's fear or panic may flow through his knees into the horse he is riding, or that emotional excitement may spread through a crowd of men without visible or audible expression. That a dog should receive a wordless message or impulse from his master at a distance of three or four miles is, fundamentally, no more unnatural than that one man should feel another's mood at a distance of three or four feet. Whether we can explain the phenomenon on strictly biological or scientific grounds is another matter.

I am not a biologist, unfortunately, and must go cat-footedly when I enter that strange garret. I look with wonder on these patient, unemotional men who care nothing for a bear or an eagle, but who creep lower and ever lower in the scale of living things, searching with penetrating looks among infinitesimal microbes for the secret that shall solve the riddle of the universe by telling us what life is. And because man is everywhere the same, watching these exploring biologists I remember the curious theology of certain South Pacific savages, who say that God made all things, the stars and the world and the living man; but we cannot see Him because He is so very small, because a dancing mote or a grain of sand is for Him a roomy palace. Yet even with a modest little knowledge of biology we may find a viewpoint, I think, from which telepathy or thought-transference would appear as natural, as inevitable, as the forthgoing of light from a burning lamp.

Thus, historically there was a time when the living cell, or the cell-of-life, as one biologist calls it with rare distinction, was sensitive only to pressure; when in its darkness it knew of an external world only by its own tremblings, in response to vibrations which poured over it from every side. Something made it tremble, and that "something" had motion or life like its own. Such, imaginatively, was the sentient cell's first knowledge, the result of a sense of touch distributed throughout its protecting surface.

Long afterward came a time when the living cell, multiplied now a millionfold, began to develop special sense-organs, each a modification of its rudimentary sense of touch; one to receive vibrations of air, for hearing; another to catch some of the thronging ether waves, for seeing; a third to register the floating particles of matter on a sensitive membrane, for taste or smelling. By that time the cell had learned beyond a peradventure that the universe outside itself had light and color and fragrance and harmony. Finally came a day when the cell, still multiplying and growing ever more complex, became conscious of a new power within itself, most marvelous of all the powers of earth, the power to think, to feel, and to be aware of a self that registered its own impressions of the external world. And then the cell knew, as surely as it knew sound or light, that the universe held consciousness also, and some infinite source of thought and feeling. Such, apparently, was the age-long process from the sentient cell to the living man.

Since we are following a different trail, this is hardly the time or place to face the question how this development from mere living to conscious life took place, even if one were wise or rash enough to grapple with the final problem of evolution. Yet it may not be amiss while we "rest a pipe," as the voyageurs say, to point out that, of the two possible answers to our question (aside from the convenient and restful answer that God made things so), only one, curiously enough, has thus far been considered by our physical scientists. The thousand books and theories of evolution which one reads are all reducible to this elementary proposition: that the simple things of life became complex by inner necessity. In other words, an eye became an eye, or an oak an oak, or a man a man, simply because each must develop according to the inner law of its being.

That may be true, though the all-compelling "inner law" is still only a vague assumption, and the mystery of its origin is untouched; but why not by outer compulsion as reasonably as by inner necessity? A cell-of-life that was constantly bombarded by moving particles of matter might be compelled to develop a sense of touch, in order to save its precious life by differentiating such particles into good and bad, or helpful and harmful. A cell over which vibrations of air and ether were continually passing might be forced for its own good to develop an ear and an eye to receive such vibrations as sound and light; and a cell over which mysterious waves of thought and emotion were ceaselessly flowing might be driven to comprehend that particular mystery by developing a thought and emotion of its own.

I do not say that this is the right answer; I mention it merely as a speculative possibility, in order to get our alleged scientific mind out of its deep rut of habit by showing that every road has two sides, though a man habitually use only one; and that Reason or Law or God, or whatever you choose to call the ultimate mainspring of life, is quite as apt to be found on one side of the road as on the other. Inner necessity is not a whit more logical or more explanatory than external force or compulsion when we face the simple fact that an animal now sees and feels in the light instead of merely existing in darkness, or that primitive cells which were dimly sentient have now become as thinking gods, knowing good and evil.

What this thought of ours is we do not know. Beyond the fact that we have it and use it, thought still remains a profound mystery. That it is a living force of some kind; that it projects itself or its waves outward, as the sun cannot but send forth his light; that it affects men as surely as gravitation or heat or the blow of a hammer affects them,—all this is reasonably clear and certain. But how thought travels; what refined mental ether conveys it outward with a speed that makes light as slow as a glacier by comparison, and with a force that sends it through walls of stone and into every darkness that the light cannot penetrate,—this and the origin of thought are questions so deep that our science has barely formulated them, much less dreamed of an answer. Yet if we once grant the simple proposition that thought is a force, that it moves inevitably from its source to its object, the conclusion is inevitable that any thinking mind should be able to send its silent message to any other mind in the universe. There is nothing in the nature of either mind or matter to preclude such a possibility; only our present habit of speech, of too much speech, prevents us from viewing it frankly.

As a purely speculative consummation, therefore, the time may come when telepathy shall appear as the natural or perfect communication among enlightened minds, and language as a temporary or evolutionary makeshift. But that beckons us away to an imaginative flight among the clouds, and on the earth at our feet is the trail we must follow.

The question why our dogs, if they have the faculty of receiving a master's message at a distance, do not use it more obviously, is one that I cannot answer. Perhaps the reason is obvious enough to some of the dogs, which have a sidelong way of coming home from their roving, as if aware they had long been wanted. Or, possibly, the difficulty lies not in the dog, but in his master. Every communication has two ends, one sending, the other receiving; and of a thousand owners there are hardly two who know how properly to handle a dog either by speech or by silence. Still again, one assumption implied in the question is that dogs or any other animals of the same kind are all alike; and that common assumption is very wide of the fact. Animals differ as widely in their instinctive faculties as men in their judgments; which partly explains why one setter readily follows his master's word or hand, or enters into his mood, while another remains hopelessly dumb or unresponsive. The telepathic faculty appears more frequently, as we shall see, among birds or animals that habitually live in flocks or herds, and I have always witnessed its most striking or impressive manifestation between a mother animal and her young, as if some prenatal influence or control were still at work.

For example, I have occasionally had the good luck to observe a she-wolf leading her pack across the white expanse of a frozen lake in winter; and at such times the cubs have a doggish impulse to run after any moving object that attracts their attention. If a youngster breaks away to rush an animal that he sees moving in the woods (once that moving animal was myself), the mother heads him instantly if he is close to her; but if he is off before she can check him by a motion of her ears or a low growl, she never wastes time or strength in chasing him. She simply holds quiet, lifts her head high, and looks steadily at the running cub. Suddenly he wavers, halts, and then, as if the look recalled him, whirls and speeds back to the pack. If the moving object be proper game afoot, the mother now goes ahead to stalk or drive it, while the pack follows stealthily behind her on either side; but if the distant object be a moose or a man, or anything else that a wolf must not meddle with, then the mother wolf trots quietly on her way without a sound, and the errant cub falls into place as if he had understood her silent command.

You may observe the same phenomenon of silent order and ready obedience nearer home, if you have patience to watch day after day at a burrow of young foxes. I have spent hours by different dens, and have repeatedly witnessed what seemed to be excellent discipline; but I have never yet heard a vixen utter a growl or cry or warning of any kind. That audible communication comes later, when the cubs begin to hunt for themselves;

and then you will often hear the mother's querulous squall or the cubs' impatient crying when they are separated in the dark woods. While the den is their home (they seldom enter it after they once roam abroad) silence is the rule, and that silence is most eloquent. For hours at a stretch the cubs romp lustily in the afternoon sunshine, some stalking imaginary mice or grass-hoppers, others challenging their mates to mock fights or mock hunting; and the most striking feature of the exercise, after you have become familiar with the fascinating little creatures, is that the old vixen, who lies apart where she can overlook the play and the neighborhood, seems to have the family under perfect control at every instant, though never a word is uttered.

That some kind of communication passes among these intel-ligent little brutes is constantly evident; but it is without voice or language. Now and then, when a cub's capers lead him too far from the den, the vixen lifts her head to look at him intently; and somehow that look has the same effect as the she-wolf's silent call; it stops the cub as if she had sent a cry or a messenger after him. If that happened once, you might overlook it as a matter of mere chance; but it happens again and again, and always in the same challenging way. The eager cub suddenly checks himself, turns as if he had heard a command, catches the vixen's look, and back he comes like a trained dog to the whistle.

As the shadows lengthen on the hillside, and the evening comes when the mother must go mousing in the distant mead-ow, she rises quietly to her feet. Instantly the play stops; the cubs gather close, their heads all upturned to the greater head that bends to them, and there they stand in mute intentness, as if the mother were speaking and the cubs listening. For a brief interval that tense scene endures, exquisitely impressive, while you strain your senses to catch its meaning. There is no sound, no warning of any kind that ears can hear. Then the cubs scam-per quickly into the burrow; the mother, without once looking back, slips away into the shadowy twilight. At the den's mouth a foxy little face appears, its nostrils twitching, its eyes following a moving shadow in the distance. When the shadow is swallowed up in the dusk the face draws back, and the wild hillside is wholly silent and deserted.

You can go home now. The vixen may be hours on her hunting, but not a cub will again show his nose until she returns and calls him. If a human mother could exercise such silent, perfect discipline, or leave the house with the certainty that four or five lively youngsters would keep out of danger or mischief as completely as young fox cubs keep out of it, raising children might more resemble "one grand sweet song" than it does at present.

So far as I have observed grown birds or beasts, the faculty of silent communication occurs most commonly among those that are gregarious or strongly social in their habits. The timber-wolves of the North are the first examples that occur to me, and also the most puzzling. They are wary brutes, so much so that those who have spent a lifetime near them will tell you that it is useless to hunt a wolf by any ordinary method; that your meeting with him is a matter of chance or rare accident; that not only has he marvelously keen ears, eyes that see in the dark, and a nose that cannot be deceived, but he can also "feel" a danger which is hidden from sight or smell or hearing. Such is the Indian verdict; and I have followed wolves often and vainly enough to have some sympathy with it.

The cunning of these animals would be uncanny if it were merely cunning; but it is naturally explained, I think, on the assumption that wolves, more than most other brutes, receive silent warnings from one another, or even from a concealed hunter, who may by his excitement send forth some kind of emotional alarm. When you are sitting quietly in the woods, and a pack of wolves pass near without noticing their one enemy, though he is in plain sight, you think that they are no more cunning than a bear or a buck; and that is true, so far as their cunning depends on what they may see or hear. Once when I was crossing a frozen lake in a snow-storm a whole pack of wolves rushed out of the nearest cover and came at me on the jump, mistaking me for a deer or some other game animal; which does not speak very highly for either their eyes or their judgment. They were the most surprised brutes in all Canada when they discovered their mistake. But when you hide with ready rifle near some venison which the same wolves have killed; when you see them break out of the woods upon the ice, running free and confident to the food which they know is awaiting them; when you see them stop suddenly, as if struck, though they cannot possibly see or smell you, and then scatter and run by separate trails to a meeting-point on another lake—well, then you may conclude, as I do, that part of a wolfs cunning lies deeper than his five senses.

Another lupine trait which first surprised and then challenged my woodcraft is this: in the winter-time, when timber-wolves commonly run in small packs, a solitary or separated wolf always seems to know where his mates are hunting or idly roving or resting in their day-bed. The pack is made up of his family relatives, younger or older, all mothered by the same she-wolf; and by some bond or attraction or silent communication he can go straight to them at any hour of the day or night, though he may not have seen them for a week, and they have wandered over countless miles of wilderness in the interim.

We may explain this fact, if such it be (I shall make it clear presently), on the simple ground that the wolves, though incurable rovers, have bounds beyond which they seldom pass; that they return on their course with more or less regularity; and that in traveling, as distinct from hunting, they always follow definite runways, like the foxes. Because of these fixed habits, a solitary wolf might remember that the pack was due in a certain region on a certain day, and by going to that region and putting his nose to the runways he could quickly pick up the fresh trail of his fellows. There is nothing occult in such a process; it is a plain matter of brain and nose.

Such an explanation sounds reasonable enough; too reasonable, in fact, since a brute probably acts more intuitively and less rationally; but it does not account for the amazing certainty of a wounded wolf when separated from his pack. He always does separate, by the way; not because the others would eat him, for that is not wolf nature, but because every stricken bird or beast seeks instinctively to be alone and quiet while his hurt is healing. I have followed with keen interest the doings of one wounded wolf that hid for at least two days and nights in a sheltered den, after which he rose from his bed and went straight as a bee's flight to where his pack had killed a buck and left plenty of venison behind them.

In this case it is possible to limit the time of the wounded wolf's seclusion, because the limping track that led from the den was but a few hours old when I found it, and the only track leading into the den was half obliterated by snow which had fallen two nights previously. How many devious miles the pack had traveled in the interim would be hard to estimate. I crossed their hunting or roaming trails at widely separate points, and once I surprised them in their day-bed; but I never found the limit of their great range. A few days later that same limping wolf left another den of his, under a windfall, and headed not for the buck, which was now frozen stiff, but for another deer which the same pack had killed in a different region, some eight or ten straight miles away, and perhaps twice that distance as wolves commonly travel.

If you contend that this wounded wolf must have known where the meat was by the howling of the pack when they killed, I grant that may be true in one case, but certainly not in the other. For by great good luck I was near the pack, following a fresh trail in the gray, breathless dawn, when the wolves killed the second deer; and there was not a sound for mortal ears to hear, not a howl or a trail cry or even a growl of any kind. They followed, killed and ate in silence, as wolves commonly do, their howling being a thing apart from their hunting. The wounded wolf was then far away, with miles of densely wooded hills and valleys between him and his pack.

Do you ask, "How was it possible to know all this?" From the story the snow told. At daybreak I had found the trail of a hunting pack, and was following it stealthily, with many a cautious dètour and look ahead, for they are unbelievably shy brutes; and so it happened that I came upon the carcass of the deer only a few minutes after the wolves had fed and roamed lazily off toward their day-bed. I followed them too eagerly, and alarmed them before I could pick the big one I wanted; whereupon they took to rough country, traveling a pace that left me hopelessly far behind. When I returned to the deer, to read how the wolves had surprised and killed their game, I noticed the fresh trail of a solitary wolf coming in at right angles to the trail of the hunting pack. It was the limper again, who had just eaten what he wanted and trailed off by himself. I followed and soon jumped him, and took after him on the lope, thinking I could run him down or at least come near enough for a revolver-shot; but that was a foolish notion. Even on three legs he whisked through the thick timber so much easier than I could run on show-shoes that I never got a second glimpse of him.

By that time I was bound to know, if possible, how the limper happened to find this second deer for his comfort; so I picked up his incoming trail and ran it clear back to his den under the windfall, from which he had come as straight as if he knew exactly where he was heading. His trail was from eastward; what little air was stirring came from the south; so that it was impossible for his nose to guide him to the meat even had he been within smelling distance, as he certainly was not. The record in the snow was as plain as any other print, and from it one might reasonably conclude that either the wolves can send forth a silent food-call, with some added information, or else that a solitary wolf may be so in touch with his pack-mates that he knows not only where they are, but also, in a general way, what they are doing.

In comparison with timber-wolves the caribou is rather a witless brute; but he, too, has his "uncanny" moods, and one who patiently follows him, with deeper interest in his *anima* than in his antlered head, finds him frequently doing some odd or puzzling thing which may indicate a perception more subtle than that of his dull eyes or keen ears or almost perfect nose. Here is one example of Megaleep's peculiar way:

I was trailing a herd of caribou one winter day on the barrens (treeless plains or bogs) of the Renous River in New Brunswick. For hours I had followed through alternate thick timber and open bog without alarming or even seeing my game. The animals were plainly on the move, perhaps changing their feeding-ground; and when Megaleep begins to wander no man can say where he will go, or where stop, or what he is likely to do next. Once, after trailing him eight or ten miles, twice jumping him, I met him head-on, coming briskly back in his own tracks, as if to see what was following him. From the trail I read that there were a dozen animals in the herd, and that one poor wounded brute lagged continually behind the others. He was going on three legs; his right forefoot, the bone above it shattered by some blundering hunter's bullet, swung helplessly as he hobbled along, leaving its pathetic record in the snow.

On a wooded slope which fell away to a chain of barrens, halting to search the trail ahead, my eye caught a motion far across the open, and through the field-glass I saw my herd for the first time, resting unsuspiciously on the farther edge of the barren, a full mile or more away. From my feet the trail led down through a dense fringe of evergreen, and then straight out across the level plain. A few of the caribou were lying down; others moved lazily in or out of the forest that shut in the barren on that side; and as I watched them two animals, yearlings undoubtedly, put their heads together for a pushing match, like domestic calves at play.

Hardly had I begun to circle the barren, keeping near the edge of it but always out of sight in the evergreens, when I ran upon a solitary caribou trail, the trail of the cripple, who had evidently wearied and turned aside to rest, perhaps knowing that his herd was near the end of its journey. A little farther on I jumped him out of a fir thicket, and watched him a moment as he hobbled deeper into the woods, heading away to the west. The course surprised me a little, for his mates were northward; and at the thought I quickly found an opening in the cover and turned my glass upon the other caribou. Already they were in wild alarm. For a brief interval they ran about confusedly, or stood tense as they searched the plain and the surrounding woods for the source of danger; then they pushed their noses out and racked away at a marvelous pace, crossing the barren diagonally toward me and smashing into the woods a short distance ahead, following a course which must soon bring them and their wounded mate together. If I were dealing with people, I might say confidently that they were bent on finding out what the alarm was about; but as I have no means of knowing the caribou motive, I can only say that the two trails ran straight as a string through the timber to a meeting-point on the edge of another barren to the westward.

If you would reasonably explain the matter, remember that these startled animals were far away from me; that the cripple and myself were both hidden from their eyes, and that I was moving upwind and silently. It was impossible that they should hear or see or smell me; yet they were on their toes a moment after the cripple started up, as if he had rung a bell for them. It was not the first time I had witnessed a herd of animals break away when, as I suspected, they had received some silent, incomprehensible warning, nor was it the last; but it was the only time when I could trace the whole process without break or question from beginning to end. And when, to test the matter to the bottom, I ran the trail of the herd back to where they had been resting, there was no track of man or beast in the surrounding woods to account for their flight.

One may explain this as a mere coincidence, which is not an explanation; or call it another example of the fact that wild animals are "queer," which is not a fact; but in my own mind every action of the caribou and all the circumstances point to a different conclusion—namely, that the fear or warning or impulse of one animal was instantly transferred to others at a distance. I think, also, that the process was not wholly unconscious or subconscious, but that one animal sent forth his warning and the others acted upon it more or less intelligently. This last is a mere assumption, however, which cannot be proved till we learn to live in an animal's skin.

It is true that the event often befalls otherwise, since you may jump one animal without alarming others of the same herd; and it is possible that the degree or quality of the alarm has something to do with its carrying power, as we feel the intense emotion of a friend more quickly than his ordinary moods. In this case the solitary caribou was tremendously startled; for I was very near, and the first intimation he had of me, or I of him, was when my snow-shoe caught on a snag and I pitched over a log almost on top of him. Yet the difficulty of drawing a conclusion from any single instance appears in this: that I have more than once stalked, killed and dressed an animal without disturbing others of his kind near at hand (it may be that no alarm was sent out, for the animal was shot before he knew the danger, and in the deep woods animals pay little attention to the sound of a rifle); and again, when I have been trying to approach a herd from leeward, I have seen them move away hurriedly, silently, suspiciously, in obedience to some warning which seemed to spread through the woods like a contagion.

The latter experience is common enough among hunters of big game, who are often at a loss to explain the sudden flight of animals that a moment ago, under precisely the same outward influences, were feeding or resting without suspicion. Thus, you may be stalking a big herd of elk, or wapiti, which are spread out loosely over half a mountainside. You are keen for the master bull with the noble antlers; nothing else interests you, more's the pity; but you soon learn that the cunning old brute is hidden somewhere in the midst of the herd, depending on the screen of cow-elk to warn him of danger to his precious skin. Waiting impatiently till this vanguard has moved aside, you attempt to worm your way nearer to the hidden bull. You are succeeding beautifully, you think, when a single cow that you overlooked begins to act uneasily. She has not seen or heard you, certainly, and the wind is still in your favor; but there she stands, like an image of suspicion, head up, looking, listening, testing the air, till she makes up her mind she would as lief be somewhere else, when without cry or grunt or warning of any kind that ears can hear she turns and glides rapidly away.

Now if you value animal lore above stuffed skins, or experience above the babble of hunting naturalists, forget the big bull and his greed-stirring antlers; scramble quickly to the highest outlook at hand, and use your eyes. No alarm has been sounded; the vast silence is unbroken; yet for some mysterious reason the whole herd is suddenly on the move. To your right, to your left, near at hand or far away, bushes quiver or jump; alert brown forms appear or vanish like shadows, all silent and all heading in the direction taken by the first sentinel. One moment there are scores of elk in sight, feeding or resting quietly; the next they are gone and the great hillside is lifeless. The thrill of that silent, moving drama has more wisdom in it, yes, and more pleasure, than the crash of your barbarous rifle or the convulsive kicking of a stricken beast that knows not why you should kill him.

Such is the experience, known to almost every elk-hunter who has learned that life is more interesting than death; and I know nothing of deer nature to explain it save this—that the whole herd has suddenly felt and understood the silent impulse to go, and has obeyed it without a question, as the young wolf or fox cub obeys the silent return call of his watchful mother.

Such impulses seem to be more common and more dependable among the whales, which have rudimentary or imperfect sense-organs, but which are nevertheless delicately sensitive to external impressions, to the approach of unseen danger, to the movements of the tiny creatures on which they feed, to changes of wind or tide and to a falling barometer, as if nature had given them a first-class feeling apparatus of some kind to make up for their poor eyes and ears. Repeatedly have I been struck by this extraordinary sensitiveness when watching the monstrous creatures feeding with the tide in one of the great bays of the Newfoundland or the Labrador coast. If I lowered a boat to approach one of them, he would disappear silently before I could ever get near enough to see clearly what he was doing. That seemed odd to me; but presently I began to notice a more puzzling thing: at the instant my whale took alarm every other whale of the same species seemed to be moved by the same impulse, sounding when the first sounded, or else turning with him to head for the open sea.

A score of times I tried the experiment, and commonly, but not invariably, with the same result. I would sight a few leviathans playing or feeding, shooting up from the deep, breaching half their length out of water to fall back with a tremendous *souse;* and through my glasses I would pick up others here or there in the same bay. Selecting a certain whale, I would glide rapidly toward him, crouching low in the dory and sculling silently by means of an oar over the stern. By some odd channel of perception (not by sight, certainly, for I kept out of the narrow range of his eye, and a whale is not supposed to smell or hear) he would invariably get wind of me and go down; and then, jumping to my feet, I would see other whales in the distance catch the instant alarm, some upending as they plunged to the deeps, others whirling seaward and forging full speed ahead.

This observation of mine is not unique, as I supposed, for later I heard it echoed as a matter of course by the whalemen. Thus, when I talked with my friend, Captain Rule, about the ways of the great creatures he had followed in the old whaling-days, he said, "The queerest habit of a whale, or of any other critter I ever fell foul of, was this: when I got my boat close enough to a sperm-whale to put an iron into him, every other sperm-whale within ten miles would turn flukes, as if he had been harpooned, too." But he added that he had not noticed the same contagion of alarm, not in the same striking or instanta-neous way, when hunting the right or Greenland whale—per-haps because the latter is, as a rule, more solitary in its habits.

Wolves and caribou and whales are far from the observation of most folk; but the winter birds in your own yard may some time give you a hint, at least, of the same mysterious transfer-ence of an impulse over wide distances. When you scatter food for them during a cold snap or after a storm (it is better not to feed them regularly, I think, especially in mild weather when their proper food is not covered with snow) your bounty is at first neglected except by the house sparrows and starlings. Unlike our native birds, these imported foreigners are easily "pauperized," seeking no food for themselves so long as you take care of them. They keep tabs on you, also, waiting patiently about the house, and soon learn what it means when you emerge from your back door on a snowy morning with a broom in one hand and a pan in the other. They are feeding greedily the moment your back is turned, and for a time they are the only birds at the table. When they have gorged themselves, for they have no manners, a few tree-sparrows and juncos flit in to eat daintily. Then suddenly the wilder birds appear—jays, chicka-dees, siskins, kinglets and, oh, welcome! a flock of bob-whites—coming from you know not where, in obedience to a summons which you have not heard. Some of these may have visited the yard in time past, and are returning to it now, hunger driven; but others you have never before met within the city limits, and a few have their accustomed dwelling in the pine woods, which are miles away. How did these hungry hermits suddenly learn that food was here?

The answer to that question is simple, and entirely "sensible" if you think only of birds that live or habitually glean in your neighborhood. Some of them saw you scatter the food, or else found it by searching, while others spied these lucky ones feeding and came quickly to join the feast. For birds that live wider afield there is also an explanation that your senses can approve, though it is probably wrong or only half right: from a distance they chanced to see wings speeding in the direction of your yard, and followed them expectantly because wings may be as eloquent as voices, the flight of a bird when he is heading for food being very different from the flight of the same bird when he is merely looking for food. But these most rare visitors, kinglets or pine-finches or grosbeaks or bob-whites, that never before entered your yard, and that would not be here now had you not thought to scatter food this morning,—at these you shake your head, calling it chance or Providence or mystery, according to your mood or disposition. To me, after observing the matter closely many times, the reasonable explanation of these rare visitors is that either wild birds know how to send forth a silent food-call or, more likely, that the excitement of feeding birds spreads powerfully outward, and is felt by other starving birds, alert and sensitive, at a distance beyond all possible range of sight or hearing. By no other hypothesis can I account for the fact that certain wild birds make their appearance in my yard at a moment when a number of other birds are eagerly feeding, and at no other time, though I watch for them from one year's end to another.

Like every other explanation, whether of stars or starlings, this also leads to a greater mystery. The distance at which such a summoning call can be felt by others must be straitly limited, else would all the starving birds of a state be flocking to my yard on certain mornings; and the force by which the silent call is projected is as unknown as the rare mental ether which bears its waves or vibrations in all directions. Yet the problem need not greatly trouble us, since the answer, when it comes, will be as natural as breathing. If silent or telepathic communication exists in nature, and I think it surely does, the mystery before us is no greater than that which daily confronts the astronomer or the wireless operator. One measures the speed of light from Orion; the other projects his finger-touch across an ocean; but neither can tell or even guess the quality of the medium by which the light or the electric wave is carried to its destination.

The Swarm Spirit

V

THIS is a chapter on the wing drill of birds, the swarming of bees, the panics and unseasonal migrations of larger or smaller beasts, and other curious phenomena in which the wild creatures of a flock or herd all act in unison, doing the same thing at the same time, as if governed by a single will rather than by individual motives. If it should turn out that the single will were expressed in a voice or cry, or even in a projected impulse, then are we again face to face with our problem of animal communication.

Of the fact of collective action there is no doubt, many naturalists having witnessed it; and there is also a strictly orthodox explanation. Thus, when you see a large flock of crows "drilling" in the spring or autumn, rising or falling or wheeling all together with marvelous precision, the ornithologists resolve the matter by saying that the many crows act as one crow because they follow a "collective impulse"; that is, because the same impulse to rise or fall or wheel seizes upon them all at precisely the same moment. And this they tell you quite simply, as if pointing out an obvious fact of natural history, when in reality they are showing you the rarest chimera that ever looked out of a vacuum.

Now the wonderful wing drill of certain birds has something in it which I cannot quite fathom or understand, not even with a miracle of collective impulse to help me; yet I have observed two characteristics of the ordered flight which may help to dispel the fog of assumption that now envelops it. The first is, that the drill is seen only when an uncommonly large number of birds of the same kind are gathered together, on a sunny day of early spring, as a rule, or in the perfection of autumn weather.

The starlings[3] furnish us an excellent example of this peculiarity. For months at a stretch you see them about the house, first in pairs, next in family groups, then in larger companies, made up, I think, of birds raised in the same neighborhood and probably all more or less related; but though you watch these companies attentively from dawn to dusk, you shall never see them going through any unusual wing drill. Then comes an hour when flocks of starlings appear on all sides, heading to a common center. They gather in trees here or there about the edges of a great field or a strip of open beach, all jabbering like the blackbirds, which they imitate in their cries, flitting about in ceaseless commotion, but apparently keeping their family or tribal organization intact. Suddenly, as at a signal, they all launch themselves toward the center of the field; the hundred companies unite in one immense flock, and presto! the drill is on. The birds are no longer individuals, but a single-minded myriad, which wheels or veers with such precision that the flash of their ten-thousand wings when they turn is like the flicker of a signal-glass in the sun.

The same characteristic of uncommon numbers holds true of the crows and, indeed, of all other species of birds, save one, that ever practise the wing drill. Wild geese when in small companies, each a family unit, have a regular and beautiful flight in harrow-shaped formation; but I have never witnessed anything like a wing drill among them save on one occasion, when a thou-

---

[3] I am speaking of starlings as they now appear in southern New England. They were brought from Europe a few years ago, and are multiplying at an alarming rate. They have formed some curious new habits here; even their voices are very different from the voices of starlings as I have heard them in Europe.

sand or more of the birds were gathered together for a few days of frolic before beginning the southern migration. Nor have I ever seen the drill among thrushes or warblers or sparrows or terns or seagulls, which sometimes gather in uncounted numbers, but which do not, apparently, have the same motive that leads crows or starlings to unite in a kind of rhythmic air-dance on periodic occasions.

A second marked characteristic of the wing drill is that it is invariably a manifestation of play or sport, and that the individual birds, though they keep the order of the play marvelously well, show in their looks and voices a suppressed emotional excitement. The drill is never seen when birds are migrating or feeding or fleeing from danger, though thousands of them may be together at such a time, but only when they assemble in a spirit of fun or exercise, and their bodily needs are satisfied, and the weather or the barometer is just right, and no enemy is near to trouble them. Whatever their motive or impulse, therefore, it is certainly not universal or even widespread among the birds, since most of them do not practise the drill; nor is it in the least like that mysterious impulse which suddenly sets all the squirrels of a region in migration, or calls the lemmings to hurry over plain and forest and mountain till they all drown themselves in the distant sea; for no sooner is the brief drill over than the companies scatter quietly, each to its own place, and the individual birds are again alert, inquisitive, well balanced, precisely as they were before.

The drill is seen at its best among the plover, I think; and, curiously enough, these are the only birds I know that practise it frequently, in small or large numbers and in all weathers. I have often watched a flock come sweeping in to my decoys, gurgling like a thousand fifes with bubbles in them; and never have I met these perfectly drilled birds, which stay with us but a few hours on their rapid journey from the far north to the far south, without renewed wonder at their wildness, their tameness, their incomprehensible ways. That you may visualize our problem before I venture an explanation, here is what you may see if you can forget your gun to observe nature with a deeper interest:

You have risen soon after midnight, called by the storm and the shrilling of passing plover, and long before daylight you are waiting for the birds on the burnt-over plain. Your "stand" is a hole in the earth, hidden by a few berry-bushes; and before you, at right angles with the course of the storm (for plover always wheel to head into the wind when they take the ground), are some scores of rudely painted decoys. As the day breaks you see against the east a motion as of wings, and your call rings out wild and clear, to be echoed on the instant. In response to your whistle the distant motion grows wildly fantastic; it begins to whirl and eddy, as if a wisp of fog were rolling swiftly down-wind; only in some mysterious fashion the fog holds together, and in it are curious flickerings. Those are plover, certainly; no other birds have that perfect unity of movement; and now, since they are looking for the source of the call they have just heard, you throw your cap in the air or wave a handkerchief to attract attention. There is an answering flash of white from the under side of their wings as the plover catch your signal and turn all at once to meet it. Here they come, driving in at terrific speed straight at you!

It is better to stop calling now, because the plover will soon see your decoys; and these birds when on the ground make no sound except a low, pulsating whistle of welcome or recall. This is uttered but seldom, and unless you can imitate it, which is not likely, your whistling will do no good. Besides, it could not pos-sibly be heard. Listen to that musical babel, and let your nerves dance to it! In all nature there is nothing to compare for utter wildness with the fluting of incoming plover.

On they come, hundreds of quivering lines, which are the thin edges of wings, moving as one to a definite goal. Their keen eyes caught the first wave of your handkerchief in the distance; and now they see their own kind on the ground, as they think, and their babel changes as they begin to talk to them. Suddenly, and so instantaneously that it makes you blink, there is a change of some kind in every quivering pair of wings. At first, in the soft light of dawn, you are sure that the plover are still coming, for you did not see them turn; but the lines grow smaller, dimmer, and you know that every bird in the flock has whirled, as if at command, and is now heading straight away. You put your fingers to your lips and send out the eery plover call again and again; but it goes unheeded in that tumult of better whistling. The quivering lines are now all blurred in one; with a final flicker they disappear below a rise of ground; the birds are gone, and you cease your vain calling. Then, when you are thinking you will never see that flock again, a cloud of wings shoot up from the plain against the horizon; they fall, wheel, rise again in marvelous flight, not as a thousand individuals but as a unit, and the lines grow larger, clearer, as the plover come sweeping back to your decoys once more.

Such is the phenomenon as I witnessed it repeatedly on the Nantucket moors, many years ago. The only way I can explain the instantaneous change of flight is by the assumption, no longer strange or untested, that from some alarmed plover on the fringe or at the center of the flock a warning impulse is sent out, and the birds all feel and obey it as one bird. That the warning is a silent one I am convinced, for it seems impossible that any peculiar whistle could be heard or understood in that wild clamor of whistling. Nor is it a satisfactory hypothesis that one bird sees the danger or suspects the quality of the decoys, and all the others copy his swift flight; for in that case there must be succession or delay or straggling in the turning, and the impression left on the eye is not of succession, but of almost perfect unity of movement.

The only other explanation of the plover's action is the one commonly found in the bird-books, to which I have already briefly referred, and which we must now examine more narrowly. It assumes that all the birds of a migrating flock are moved not by individual wills, but by a collective impulse or instinct, which affects them all alike at the same instant. In support of this favorite theory we are told to consider the bees, which are said to have no individual motives, and no need for them, since they blindly follow a swarm or hive instinct that makes them all precisely alike in their actions. The same swarm instinct appears often in the birds, but less strongly, because they are more highly developed creatures, with more need and therefore more capacity for individual incentive.

This illustration of the hive is offered so confidently and accepted so readily, as if it were an axiom of natural history, that one hesitates to disturb the ancient idol in its wonted seat. Yet one might argue that any living impulse, whether in bees or birds, must proceed from a living source, and, if that be granted, speculate on the absorbing business of a nature or a heaven that should be perpetually interfering in behalf of every earthly flock or swarm or herd by sending the appropriate impulse at precisely the right moment. And when our speculation is at an end, I submit the fact that, when I have broken open a honey-tree in the woods, one bee falls upon the sweets to gorge himself withal, while another from the same swarm falls angrily upon me and dies fighting; which seems to upset the collective-impulse idol completely.

I must confess here that I know very little about bees. They are still a mystery to me, and I would rather keep silence about them until I find one bee that I fancy I understand, or one man who offers something better than a very hazy or mystical explanation of a bee's extraordinary action. Yet I have watched long hours at a hive, have handled a swarm without gloves or mask, and have performed a few experiments—enough to convince me that the collective-impulse theory does not always hold true to fact even among our honey-makers. Indeed, I doubt that it ever holds true, or that there is in nature any such mysterious thing as a swarm or flock or herd impulse.

In the first place, the bees of the same swarm do not look alike or act alike except superficially; at least I have not so observed them. Study the heads or the feet of any two bees under a glass, and you shall find as much variety as in the heads or feet of any other two creatures of the same kind, whether brute or human. The lines of difference run smaller, to be sure; but they are always there. In action also the bees are variable; they do marvelously wise things at one moment, or marvelously stupid things at another; but they do not all and always do the same thing under the same circumstances, for when I have experimented with selected bees from the same hive I have noticed very different results; which leads me to suspect that even here I am dealing with individuals rather than with detached fragments of a swarm. It is hard, for example, to make a trap so simple that an imprisoned bee will find his way out of it; but when by great ingenuity you do at last make a trap so very simple that it seems any creature with legs must walk out by the open door, perhaps one bee in five will do the trick; while the other four wait patiently until they die for more simplicity.

Again, while your eye often sees unity of action among the wild creatures, neither your reading nor your own reason will ever reveal a scrap of positive evidence that there is in nature any such convenient thing (humanly convenient, that is, for explanations) as a swarm or flock instinct; though, like the mythical struggle for existence, we are forever hearing about it or building theories upon it. So far as we know anything about instinct, it is neither collective nor incorporeal. It is, to use the definition of Mark Hopkins, which is as good as another and beautifully memorable, "a propensity prior to experience and independent of instruction." And the only needful addition to this high-sounding definition is, that it is a "propensity" lodged in an individual, every time. It is not and cannot be lodged in a swarm or a hive; you must either put it into each of two bees or else put it between them, leaving them both untouched. In other words, the swarm instinct has logically no abiding-place and no reality; it is a castle in the air with no solid foundation to rest on.

On its practical or pragmatic side also the theory is a failure, since the things bees are said to do in obedience to an incorporeal swarm instinct are more naturally and more reasonably explained by other causes. Bees swarm, apparently, in the lead or under the influence of individuals; and it needs only a pair of eyes to discover that there are plenty of individual laggards and blunderers in the process. They grow angry not all at once, but successively; not because a swarm instinct impels them to anger, but because one irritated bee gives off a pungent odor or raises a militant buzzing, and the others smell the odor or hear the buzzing and are inflamed by it, each through his own senses and by the working of his own motives. On a hot day you will see a few bees fanning air into the hive with their wings, and when these grow weary others take their places; but if it were a swarm instinct that impelled them, you would see all the bees fanning or all sweltering at the same moment. As for the honey-making instinct, on any early-spring day you will find a few bees working in the nearest greenhouse, while the others, which are supposed to be governed by the same collective impulse, are comfortably torpid in the hive or else eating honey faster than these enterprising ones can make it.

I judge, therefore, that the communistic bees have some individual notions, and any show of individuality is so at variance with the common-impulse theory that it seems to illustrate Spencer's definition of tragedy, which is, "a theory slain by a fact." In short, bees have our common social instinct highly developed, or overdeveloped, and possibly they have also, like all the higher orders, a stronger or weaker instinct of imitation; but these are very different matters, more natural and more consistent with the facts than is the alleged swarm instinct.

A scientific friend, the most observant ornithologist I have ever met, has just offered an interesting explanation of the flock or herd phenomena we are here considering. He finds little evidence of a swarm instinct, as distinct from our familiar social instinct; but he has often marveled at the wing drill of birds, and has twice witnessed an alarm or warning of danger spread silently among a herd of scattered beasts; and he accounts for the observed facts by the supposition that the minds (or what corresponds to the minds) of the lower orders are often moved not from within, but from without—that is, not by instinct or by sense impressions, not by what they or others of their kind may see or hear, but by some external and unknown influence. My caribou rushed away, he thinks, and my incoming plover turned as one bird from my decoys, because a warning impulse fell upon them at a moment when they were in danger, but knew it not; and they obeyed it, as they obey all their impulses, without conscious thought or knowledge of what they are doing or why they are doing it.

Here is some suggestion of a very modern psychology which is inclined to regard the mind as a thought-receiving rather than as a thought-producing instrument, and with that I have some sympathy; but here is also a rejuvenation of the incorporeal swarm instinct and other fantastic or romantic notions of animals which preclude observation. If the *anima* of a bird or beast is so constituted that it can receive impulses from a mysterious and unknown source, what is to prevent it from receiving such silent impulses from another *anima* like itself? And why seek an unseen agent for the warning to my caribou or my plover when one of the creatures saw the danger and was enough moved by it to sound a mental tocsin?

The trouble with my friend's explanation, and with all others I have thus far heard or read, is twofold. First, like the swarm-impulse theory, it really explains nothing, but avoids one mystery or difficulty by taking refuge in another. There was a Hindu philosopher who used to teach, after the manner of his school, that the earth stood fast because it rested on the back of a great elephant; which was satisfactory till a thoughtful child asked, "But the elephant, what does he stand on?" So when I see intelligent caribou or plover fleeing from an unsensed danger, and am told that they have received an impulse from without, I am bound to ask, "Where did that impulse come from, and who sent it?" For emotional impulses do not drop like rain from the clouds, or fall like apples from unseen trees; they must have their source in a living, intelligent being of some kind, who must feel the impulse before sending it to others. No other explanation is humanly comprehensible.

This leads to the second objection to the theory of external impulse, and to every other notion of a collective or incorporeal swarm instinct—namely, that it contradicts all the previous experience of the wild creature, or at least all educative experience, which lies plain and clear to our observation. To each bird and animal are given individual senses, individual wit and a personal *anima;* and each begins his mortal experience not in a great flock or herd, but always in solitary fashion, under the care and guidance of a mother animal that has a saving knowledge of a world in which the little one is a stranger. Thus, I watch the innocent fawn when it begins to follow the wary old doe, or the fledgling snipe as it leaves the nest under expert guidance, or the wonder-eyed cub coming forth from its den at the call of the gaunt old she-wolf. In each case I see a mother intelligently caring for her young, leading them to food, warding them from danger, calling them now to assemble or now to scatter; and before my eyes these ignorant youngsters quickly learn to adapt themselves to the mother's ways and to obey her every signal. Sometimes I see them plainly when some manner of silent communication passes among them (something perhaps akin to that which passes when you catch a friend's eye and send your thought or order to him across a crowded room), and it has even seemed to me, as recorded elsewhere in our observation of wolf and fox dens, that the young understand this silent communication more readily than they learn the meaning of audible cries expressive of food or danger.

Such is the wild creature's earliest experience, his training to accommodate himself to the world, and to ways that wiser creatures of his own kind have found good in the world. When his first winter draws near he is led by his mother to join the herd or pack or migrating flock; and he is then ready not for some mysterious new herd or flock instinct, but for the same old signals that have served well to guide or warn him ever since he was born. I conclude, therefore, naturally and reasonably, that my caribou broke away and my incoming plover changed their flight because one of their number detected danger and sent forth a warning impulse, which the others obeyed promptly because they were accustomed to just such communications. There was nothing unnatural or mysterious or even new in the experience. So far as I can see or judge, there is no place or need for a collective herd or flock impulse, and the birds and beasts have no training or experience by which to interpret such an impulse if it fell upon them out of heaven.

Our human experience, moreover, especially that which befalls on the borderland of the subconscious world where the wild creatures mostly live, may give point and meaning to our natural philosophy. There are emotions, desires, impulses which may be conveyed by shouting; and there are others which may well be told without shouting, or even without words. A cheerful man radiates cheerfulness; a strong man, strength; a brave man, courage (we do not know to what extent or with what limitations); and a woman may be more irritated by a man who says nothing than by a man who says too much. These common daily trials may be as side-lights on the tremendous fact that love, fear, hate,—every intense emotion is a force in itself, a force to be reckoned with, apart from the cry or the look by which it is expressed; that all such emotions project themselves outward; and that possibly, or very probably, there is some definite medium to convey them, as an unknown medium which we call "ether" conveys the waves of light.

It is true that we habitually receive such emotional impulses from others by means of our eyes or ears; but sometimes we apparently imbibe them through our skin, as Anthony Trollope said he learned Latin, and once in a way we receive them from another without knowing or thinking of the process at all. It is noteworthy that the most companionable people in the world are silent people, especially a silent friend, and that the silence of any man is invariably more eloquent than his speech. The silence of one man rests you like a melody; the silence of another bores you to yawning, perhaps because it is a "dead" silence; the quietude of a third excites your curiosity to such an extent that, for once in your life, you behave like a perfectly natural animal; that is, you go round the silent one, as it were, view him mentally from all sides, sniff at his opinions from leeward, whir your wings in his face like a sparrow, or stamp your foot at him like a rabbit—all this to stir him up and to uncover what interesting thing lies behind his silence. And why? Simply because every living man is silently, unconsciously projecting his real thought or feeling, and you are unconsciously understanding it or else making a vain conscious effort in that direction.

Such experiences are commonly confined to a room, to the circle of an open fire; but they are not limited by necessity to any narrow reach, since there is nothing in a wall to hinder a man's love or hate from passing through, or in the air to check its far-going, or in the nature of another man to prevent its reception. The influence of one person's unvoiced will or purpose or warning or summons upon another person at a distance, should it turn out more common than we now believe possible because of our habit of speech, would be nothing unnatural or mysterious, but rather a true working of the subconscious or animal mind, which had its own way of communication before ever speech was invented.

Whitman, who sometimes got hold of the tail end of philosophy (and who was wont to believe he could drag it out, like a trapped woodchuck, and whirl it around his head with barbaric whoops), was often seen at the burrow of this thought-transference doctrine:

These yearnings why are they? these thoughts in the darkness why are they?
Why are there men and women that while they are near me the sun-light expands my blood?
Why when they leave me do my pennants of joy sink flat and lank?
Why are there trees I never walk under but large and melodious thoughts descend upon me?
What is it that I interchange so suddenly with strangers?
What with some driver as I ride on the seat by his side?
What with some fisherman drawing his seine by the shore as I walk by and pause?
What gives me to be free to woman's and man's good-will? what gives them to be free to mine?

Again, our familiar human experience may throw some clear-er light than ever comes from the laboratory of animal psychol-ogists upon the action of gregarious brutes in their so-called blind panics, when they are said to be governed by some extra-neous or non-individual herd impulse. How such a theory origi-nated is a puzzle to one who has closely observed animals in the open, since their panics are never "blind," and their "extrane-ous" impulse may often be traced to an alarmed animal of their own kind, or even to an excited human being, whose emotions are animal-like both in their manifestation and in their irritating effect. A dog is more easily roused by human than by canine excitement. A frightened rider sends his fear or irresolution in exaggerated form into the horse beneath him. The herd of swine that ran down a steep place into the sea were possessed, I should say, not by exorcised demons, but by the hysteria received directly from some man or woman of the excited crowd in the immediate neighborhood. Panic is more infectious than any fever, and knows no barriers between brute and human.

Indeed, in a frightened crowd in the Subway, in a theater where smoke appears, or in any other scene of emotional excitement, you may in a few minutes observe actions more panicky, more suggestive of a herd impulse (if there be such a fantastic thing in orderly nature), than can be seen in a whole lifetime of watching wild animals.

In my head at this moment is the vivid impression of a night when I was caught and carried away by a crowd of Italian socialists, twenty thousand frenzied men and a few ferocious women, that first eddied like a storm-tide about the great square under the cathedral at Milan, howling, shrieking, imprecating, and then poured tumultuously through choked streets to hurl paving-blocks at the innocent roof of the railroad station, as at a symbol of government. The roof was of glass, and the clattering smash of it seemed to get on the nerves of men, like the cry of *sick-em!* to an excited dog, rousing them to a senseless fury of destruction. Clear and thrilling above the tumult a bugle sang, like a note from heaven, and into the seething mass of humanity charged a squadron of cavalry, striking left or right with the flats of their sabers, raising a new hubbub of shrieks and imprecations as the weaker were trampled down. Fear? That crowd knew no more of fear just then than an upturned hive of bees. They met the charge with a roar, a hoarse, solid shout that seemed to sweep the cavalry away like smoke in the wind. Unarmed men swarmed at the horses like enraged baboons, hurling stones or curses as they went. The rush ended in a triumphant yell, and riderless horses, their eyes and nostrils aflame, went plunging, kicking, squealing through the pandemonium.

There must have been something tremendously *animal* in the scene, after all; for when I recall it now I see, as if Memory had carved her statue of the event, an upreared horse with a crumpled rider toppling from the saddle; and I hear not the shouts or curses of men, but the horrible scream of a maddened brute.

It was the night, many years ago, when news of disaster to the Italian army at Adowa broke loose, after being long suppressed, and I learned then for the first time what emotional excitement means when the gates are all down. One had to hold himself

against it, as against a flood or a mighty wind. To yield, to lose
self-control even for an instant, was to find oneself howling,
reaching for paving-blocks, seeking an enemy, lifting a bare fist
against charging horse or swinging steel, like the other lunatics.
I caught a man by the shoulders, held him, and bade him in his
own language tell me what the row was about; but he only stared
at me wildly, his mouth open. I caught another, and he struck at
my face; a third, and he shrieked like a trapped beast. Only one
gave me a half-coherent answer, a man whom I dragged from
under a saber and pushed into a side-street. His dear Ambrogio
had been conscripted by the government, he howled (I suppose
they had sent his son or brother with a disaffected Milanese
regiment on the African adventure), and they were all robbers,
oppressors, murderers—he finished by jerking loose from my
grasp and hurling himself, yelling, into the mob again.

Had I been a visiting caribou, watching that amazing scene
and knowing nothing of its motive, I might easily have concluded that some mysterious herd impulse was driving all these
creatures to they knew not what; but, being human, I knew perfectly well that even this unmanageable crowd had taken its cue
from some leader; that the senseless emotion which inflamed
them had originated with individuals, who had some ground for
their passion; and that from the individual the excitement
spread in pestilential fashion until the whole mob caught it and
bent to it, as a field of grass bends to the storm.

Therefore (and I hope you keep the thread of logic through a
long digression), when I go as a man among caribou or wolves
or plover or crows, and see the whole herd or pack or flock acting as one, as if swayed by a single will, I see no reason why I
should evoke an incorporeal swarm impulse, or "call a spirit
from the vasty deep" of the unknown to explain their similarity
of action, since there are natural causes which may account for
the matter perfectly—familiar causes, too, which still influence
men and women as they influence the remote wood folk.

No, this is not a new animal psychology; it is rather an attempt
to banish the delusion that there is any such thing as a distinct
animal psychology. Science has many forms, and still plenty of
delusions, but there is a basic principle to which she holds

steadily—namely, that Nature is of one piece because her laws are constant. It follows that, if you know anything of a surety about your own mind, you may confidently apply the knowledge to any other mind in the universe, whether in the heavens above or the earth beneath or the waters under the earth. The only question is, How far may the term "mind" be properly applied to the brute?

That unanswered question does not immediately concern us, for in speaking of mind we commonly mean the conscious or reasoning human article, and we are dealing here with the subconscious mind, which seems to work after the same fashion whether it appears on two legs or four. A dog does not know *why* he becomes excited in a commotion that does not personally concern him, or why he feels impelled to hasten to an outcry from an unknown source, or why he looks up, contrary to all his habits, when everybody else is looking up; and neither does a man know why he does just such things. Man and brute both act in obedience to something deeper, more primal and more dependable than reason, and in this subconscious field they are akin; otherwise it would be impossible for a man ever to train or to understand a brute, and our companionable dogs would be as distant as the seraphim.

When, therefore, the same unreasoning actions that are attributed to a mysterious collective impulse among birds or animals are found among men to depend on a succession of individual impulses, it is good psychology as well as good natural history to dismiss the whole herd instinct as another thoughtless myth. The familiar social and imitative instincts, the contagion of excitement, the outward projection of emotional impulses, the sensitive bodily nature of an animal which enables him to respond to such impulses even when they are unaccompanied by a voice or cry,—these are comparatively simple and "sensible" matters which explain all the phenomena of flock or herd life more naturally and more reasonably.

Where Silence is Eloquent

VI

LOOKING back a moment on our trail of animal "talk" before following it onward, we see, first, that birds and beasts have certain audible cries which convey a more or less definite meaning of food or danger or assembly; and second, that they apparently have also some "telepathic" faculty of sending emotional impulses to others of their kind at a distance. The last has not been proved, to be sure; we have seen little more than enough to establish it as a working hypothesis; but whether we study science or history or an individual bird or beast, it is better to follow some integrating method or principle than to blunder around in a chaos of unrelated details. And the hypothesis of silent communication certainly "works," since it helps greatly to clarify certain observed phenomena of animal life that are otherwise darkly mysterious.

When the same dimly defined telepathic power appears in a man or woman—so rarely that we are filled with wonder, as in the shadow of a great mystery or a great discovery—it is not a new but a very old matter, I think, being merely a survival or reappearance of a faculty that may have once been in common use among gregarious creatures. All men seem to have some hint or suggestion of telepathy in them, as shown by their ability to "speak with their eyes" or to influence their children by a look; and the few who have enough of it to be conspicuous receive it, undoubtedly, by some law or freak of heredity, such as enables one man in a million to wag his ears, or one in a thousand to follow a subconscious sense of direction so confidently that, after wandering about the big woods all day, he turns at nightfall and heads straight for his camp like a homing pigeon. The rest of us, meanwhile, by employing speech exclusively to express thought or emotion, and by habitually depending on five senses for all our impressions of the external world, have not only neglected but even lost all memory of the gift that once was ours. As an inevitable consequence Nature has taken her gift away, as she atrophies a muscle that is no longer used, or devitalizes the nerve of sight in creatures, such as the fishes of Mammoth Cave, that have lived long time in darkness.

In previous chapters we have noted, as rare examples of telepathy in human society, that a mother may at times know when an absent son or husband is in danger, or that an African savage often knows when a stranger is approaching his village hid in the jungle; but there is another manifestation of the same faculty which is much more common, and which we have thus far overlooked, leaving it as an odd and totally unrelated thing without explanation. I refer to the man, known in almost every village, who has some special gift for training or managing animals, who seems to know instinctively what goes on in a brute's head, and who can send his own will or impulse into the lower mind. I would explain that unrelated man, naturally, by the simple fact or assumption that he has inherited more than usual of the animals' gift of silent communication.

I knew one such man, a harmless, half-witted creature, who loved to roam the woods alone by day or night, and whom the wild birds and beasts met with hardly a trace of the fear or suspicion they manifest in presence of other human beings. He was always friendly, peaceable, childlike, and unconsciously or subconsciously, I think, he could tame or influence these wild spirits by letting them feel his own.

So also could an old negro, an ex-slave, with whom I used to go fox-hunting in my student days. He could train a dog or a colt in a tenth part of the time required by ordinary men, and he used no whip or petting or feeding, or any other device commonly employed by professional trainers. At times, indeed, his animals acted as if trained from the moment he touched or spoke to them. He had a mongrel lot of dogs, cats, chickens, pigs, cows and horses; but they were a veritable happy family (on a cold night his cats would sleep with a setting hen, if they could find one, or otherwise with the foxhounds), and to see them all running to meet "Uncle" when he came home, or following at his heels or doing what he told them, was to wonder what strange animal language he was master of.

At daybreak one winter morning I entered the old negro's kitchen very quietly, and had a fire going and coffee sending forth its aroma before I heard him creaking down the stairs. I had traveled "across lots," making no sound in the new-fallen snow, and, as I approached the house, had purposely kept its dark bulk between me and the dogs, which were asleep in their kennel some distance away. For a time all was quiet as only a winter dawn can be; but as we sat down to breakfast one of the hounds with a big bass voice suddenly broke out in an earth-shaking jubilation. The other hounds quickly caught up the clamor, yelling as if they had just jumped a fox, while two dogs of another breed were strangely silent; and then 'Poleon added his bit to the tumult by stamping, whinnying and finally kicking lustily on the boards of his stall. 'Poleon, by the way, was an old white horse that Uncle used to ride (he was "gittin' too rheumaticky," he said, to hunt with me afoot), and this sober beast was fair crazy to join the chase whenever a fox was afoot.

The negro paid no attention to the noise; but as it went on increasing, and 'Poleon whinnyed more wildly, and the big-voiced hound kept up a continuous bellow that might have roused the seven sleepers, the unseemly racket got on my nerves, so early in the day.

"What the mischief is the matter with Jum this morning?" I demanded.

"Matter? Mischief?" echoed Uncle, as if surprised I did not

understand such plain animal talk. "Why, ol' Jum's a-gwine fox-huntin' dis mawny. He reckons he knows what we-all's up to: and now de yother dawgs an' 'Poleon dey reckons dey knows it, too. Jum's tole um. Dat's all de matter an' de mischief."

"But how in the world should he know? You never go hunting now unless I tempt you, and none of the dogs saw or heard me come in," I objected.

Uncle chuckled at that, chuckled a long time, as if it were a good joke. "Trust ol' Jum ter know when we-all's gwine fox-huntin'," he said. "You jes' trust *him*. I specks he kinder pick de idee outer de air soon's we thunk it, same's he pick a fox scent. 'Tain't no use tryin' ter lie ter Jum, 'cause you can't fool 'im nohow. No, sir, when dat ol' dawg's eroun', you don' wanter think erbout nothin' you don' want 'im ter know."

I had often marveled at Uncle, but now suddenly I thought I understood him. In his unconscious confession that he thought or felt with his animals, rather than spoke English to them, was probably the whole secret of his wonderful gift of training.

The same "secret" is shared by the few men who have the gift of managing horses, and who can do more by a word or even a look than another man by bit and harness. I have heard the gift described by a professional horse-trainer as the "power of the human eye"; but that is nonsense set to melodrama. An eye is a bit of jelly, and means nothing without a will or communicating impulse behind it. When the spirit of a horse is once broken (and most of them are broken rather than trained by our methods) almost anybody can manage him, the blind as well as the seeing; but when a horse keeps the spirit of his wild ancestors, their timidity, their flightiness, their hair-trigger tendency to shy or to bolt, then I envy the man who can cross the gulf of ages and put something of his own steadiness into the nervous brute.

This steadying process seems to be wholly a matter of spirit, so far as I have observed it, and whatever passes from man to brute passes directly, without need of audible speech. For example, a friend of mine, a very quiet man of few words, once brought home a magnificent "blooded" horse which he had bought for a song because "nobody could handle him." The horse was not vicious in any way, but seemed to have a crazy impulse to run himself to death—an impulse so strong that even now, when he is past twenty years old, he cannot be turned loose for a moment in a farm pasture. He had never been driven save with a powerful curb; even so, he would drag the carriage along by the reins, and an hour of such driving left a strong man's arms half paralyzed by the strain. Yet at the first trial his new owner put a soft rubber bit in his mouth, flipped the lines loosely across his back, and controlled him by a word.

Some years later I was riding behind that same horse, jogging quietly along a country road, when my friend, with an odd twinkle in his eye, said, "Take the reins a moment while I get out this robe." I took them, and what followed seemed like magic or bedevilment. I had noticed that the reins were loose, just "feeling" the horse's mouth; I shifted them to my hand very quietly, without stirring a hair, and blinders on the bridle prevented the horse from seeing the transfer. Yet hardly had I touched them when something from my hand (or from my soul, for aught I know) flowed along the leather and filled the brute with fire. He flung up his head, as if I had driven spurs into him, and was away like a shot.

Again, I was crossing the public square of Nantucket one morning when I saw a crowd of excited men and boys eddying at a safe distance around a horse—an ugly, biting brute that had once almost torn the side of my face off when I passed too close to him, minding my own affairs. Now he was having one of his regular tantrums, squealing, kicking, plunging or backing, while his driver, who had leaped to the ground, alternately lashed and cursed him. I heard an angry voice near me utter the single word "Fools!" and saw a stranger brush some men aside and stand directly in front of the horse, which grew quiet on the instant. The stranger went nearer, pulled the horse's head down and laid his face against it; and there they stood, man and brute, like carved statues. It was as if one were whispering a secret, and the other listening. Then the man said, "Come along, boy," and walked down the square, the horse following at his heels like a trained dog.

Watching the scene, my first thought was that the horse recognized a former and kinder master; but the man assured me, when I followed him up, that he had never spoken to the animal till that moment, and that he could do the same with any refractory horse he had ever met. "Try it with that one," I said promptly, pointing to a nervous horse that, feeling the excitement of the recent affair, was jerking and frothing at his hitch-rope. The man smiled his acceptance of the challenge, stepped in front of the horse, and looked at him steadily. What he thought or willed, what feeling or impulse he sent out, I know not; but certainly some silent communication passed, which the horse recognized by forward-pointing ears and a low whinny of pleasure. Then the man unsnapped the rope from the bridle ring, turned away without a word, and the horse followed him across the square and back again to the hitching-post.

When I asked how the thing was done, the man answered with entire frankness that he did not know. It "just came natural" to him, he said, to understand horses, and he had always been able to make them do almost anything he wanted. But he had no remarkable power over other animals, so far as I could learn, and was uncommonly shy of dogs, even of little dogs, regarding them all alike as worthless or dangerous brutes.

Some of my readers may recall, in this connection, the shabby-genteel old man who used to amuse visitors in the public gardens of Paris by playing with the sparrows, some twenty-odd years ago. So long as he went his way quietly the birds paid no more attention to him than to any other stroller; but the moment he began to chirp some wild and joyous excitement spread through the trees. From all sides the sparrows rushed to him, alighting on his hat or shoulders, clamoring loudly for the food which they seemed to know was in his pockets, but which he would not at first give them. When he had a crowd of men and women watching him (for he was vain of his gift, and made a small living by passing his hat after an entertainment) he would single out a cock-sparrow from the flock and cry, "What! *you* here again, Bismarck, you scoundrel?" Then he would abuse the cock-sparrow, calling him a barbarian, a Prussian, a mannerless intruder who had no business among honorable French sparrows; and finally, pretending to grow violently angry, he would chase Bismarck from bench to bench and throw his hat at him. And Bismarck would respond by dodging the hat, chirping blithely the while, as if it were a good joke, and would fly back to peck at the crust of bread which the old man held between his lips or left sticking out of his pocket.

One might have understood this as a mere training trick if Bismarck were always the same; but he was any cock-sparrow that the man chanced to pick out of a flock. After playing with the birds till they wearied of it, he would feed them, pass the hat, and stroll away to repeat his performance with another flock in another part of the gardens. That these wary and suspicious birds, far more distrustful of man than the sparrows of the wilderness, understood his mental attitude rather than his word or action; that he could make them feel his kindliness, his *camaraderie,* his call to come and play, even while he pretended to chase them,—this was the impression of at least one visitor who watched him again and again at his original entertainment.

Some kind of communication must have passed silently between the actor and one of his audience; for presently, though I never spoke to the old man, but only watched him keenly, he picked me out for personal attention. Whereupon I cultivated his acquaintance, invited him to dine and fed him like a duke, and thought I had gained his confidence by taking him to see a big wolf of mine that might well have puzzled any student of birds or beasts. The wolf was one of a wild pack that had recently arrived at the zoo from Siberia, where they had been caught in a pit and shipped away with all their savagery in them. Through some freak of nature this one wolf had attached himself to me, like a lost dog; by some marvelously subtle perception he would recognize my coming at a distance, even in a holiday crowd, and would thrust his grim muzzle against the bars of his cage to howl or roar till I came and stretched out a hand to him, though he was as wild and "slinky" as the rest of the pack to everybody else, even to the keeper who fed him. That interested the sparrow-tamer, of course; but he was silent or too garrulous whenever I approached the thing I wanted to know. He would not tell me how he won the birds, but made a mystery and hocus-pocus of the natural gift by which he earned a precarious living.

The same "mystery" cropped out later, amid very different scenes, in the interior of Newfoundland. Coming down beside a salmon river one day, my Indian, a remarkable man with an almost uncanny power of calling wild creatures of every kind, pointed to a hole high on the side of a stub, and said, "Go, knock-um dere; see if woodpecker at home." I went and knocked softly, but nothing happened. "Knock-um again, knock-um little louder," said Matty. I knocked again, more lustily, and again nothing happened. Then the Indian came and rapped the tree with his knuckles, while I stood aside; and instantly a woodpecker that was brooding her eggs stuck her head out of the hole and looked down at her visitor inquisitively.

The next day at the same place we repeated the same performance precisely, after our morning fishing; and again the interesting thing to me was, not the bird's instant appearance at the Indian's summons, but the curiously intent way in which she turned her head to look down at him. When he showed his craft again and again, at the doors of other woodpeckers that were not interested in my knocking, I demanded, "Now, Matty, tell me how you do it."

But Matty only laughed. When we are alone in the woods he has a fine sense of humor, though grim enough at other times. "Oh, woodpecker know me; he look down at me," he said; and that was all I could ever get out of him.

So, though I have seen the gift in operation several times, I have not yet found the man who had it and who could or would give me any explanation. There is no doubt in my own mind, however, that the negro, the Frenchman, and the Indian, and all others who exercise any unusual influence over animals, do so by reason of their subconscious power to "talk" or to convey impulses without words, as gregarious wild creatures commonly talk among themselves. At least, I can understand much of what I see among birds and brutes by assuming that they talk in this fashion.

Such a power seems mysterious, incredible, in a civilized

world of sense and noise; but I fancy that every man and woman speaks silently to the brute without being conscious of the fact. "If you want to see game, leave your gun at home," is an accepted saying among hunters; but the reason for the excellent admonition has not been forthcoming. When you have hunted six days in vain, and then on a quiet Sunday stroll come plump upon noble game that seems to have no fear, you are apt to think of the curiosities of luck, but even here also are you under the sway of psychological law and order. As you go quietly through the woods, projecting your own "aura" of peace or sympathy, it may be, in an invisible wave ahead of you, there is nothing disturbing or inharmonious in either your thoughts or your actions; and at times the wild animal seems curiously able to understand the one as well as the other, just as certain dogs know at first glance whether a stranger is friendly or hostile or afraid of them. When you are excited or lustful to kill, something of your emotional excitement seems to precede you; it passes over many wild birds and beasts, all delicately sensitive, before you come within their sense range; and when you draw near enough to see them you often find them restless, suspicious, though as yet no tangible warning of danger has floated through the still woods. At the first glimpse or smell of you they bound away, your action in hiding or creeping making evident the danger which thus far was only vaguely felt. But if you approach the same animals gently, without mental excitement of any kind, sometimes, indeed, they promptly run away, especially in a much-hunted region; but more frequently they meet you with a look of surprise; they move alertly here or there to get a better view of you, and show many fascinating signs of curiosity before they glide away, looking back as they go.

Such has been the illuminating experience of one man, at least, repeated a hundred times in the wilderness. I have been deep in the woods when my food-supply ran low, or was lost in the rapids, or went to feed an uninvited bear, and it was then a question of shoot game or go hungry; but the shooting was limited by the principle that a wild animal has certain rights which a man is bound to respect. I have always held, for example, that a hunter has no excuse for trying long shots that are beyond his ordinary skill; that it is unpardonable of him to "take a chance" with noble game or to "pump lead" after it, knowing as he does that the chances are fifty to one that, if he hits at all, he will merely wound the animal and deprive it of that gladness of freedom which is more to it than life. So when I have occasionally gone out to kill a buck (a proceeding which I heartily dislike) I have sometimes hunted for days before getting within close range of the animal I wanted. But when, in the same region and following the same trails, I have entered the big woods with no other object than to enjoy their stillness, their fragrance, their benediction, it is seldom that I do not find plenty of deer, or that I cannot get as near as I please to any one of them. More than once in the woods I have touched a wild deer with my hand (as recorded in another chapter) and many times I have had them within reach of my fishing-rod.

It is even so with bear, moose, caribou and other creatures— your best "shots" come when you are not expecting them, and it is not chance, but psychological law, which determines that you shall see most game when you leave your gun at home. A hunter must be dull indeed not to have discovered that the animal he approaches peaceably, trying to make his eyes or his heart say friendly things, is a very different animal from the one he stalks with muscles tense and eyes hard and death in the curl of his trigger finger.

I once met an English hunter, a forest officer in India, who told me that for the first year of his stay in the jungle he was "crazy" to kill a tiger. He dreamed of the creatures by night; he hunted them at every opportunity and in every known fashion by day; he never went abroad on forest business without a ready rifle; and in all that time he had just one glimpse of a running tiger. One day he was led far from his camp by a new bird, and as he watched it in a little opening, unarmed and happy in his discovery, a tiger lifted its huge head from the grass, not twenty steps away. The brute looked at him steadily for a few moments, then moved quietly aside, stopped for another look, and leaped for cover.

That put a new idea into the man's head, and the idea was emphasized by the fact that the unarmed natives, who had no desire to meet a tiger, were frequently seeing the brutes in regions where he hunted for them in vain. As an experiment he left his rifle at home for a few months; he practised slipping quietly through the jungle without physical or mental excitement, as the natives go, and presently he, too, began to meet tigers. In one district he came close to four in as many months, and every one acted in the same half-astonished, half-inquisitive way. Then, thinking he understood his game, he began to carry his rifle again, and had what he called excellent luck. The beautiful tiger skins he showed me were a proof of it.

To me this man was a rare curiosity, being the only Indian or African hunter I ever met who went into the jungle alone, man fashion, and who did not depend on unarmed natives or beaters or trackers for finding his game. His excellent "luck" was, as I judge, simply a realization of the fact that human excitement may carry far in the still woods, and be quite as disturbing, as the man-scent or the report of a rifle.

Does all this sound strange or incredible to you, like a chapter from a dream-book? However it may sound, it is the crystallized conviction resulting from years of intimate observation of wild beasts in their native woods; and if you consider it a moment without prejudice, it may appear more natural or familiar, like a chapter from life. If the man who sits opposite you can send his good or evil will across a room, so that you feel his quality without words, or if he can so express himself silently when he enters your gate that certain dogs instantly take his measure and welcome or bite him, it is not at all improbable that the same man can project the same feelings when he goes afield, or that sensitive wild creatures can understand or "feel him out" at a considerable distance.

To weigh that probability fairly you must first get rid of your ancient hunting lore. Hunters are like the Medes and Persians in that they have laws which alter not; and I suppose if you met ancient Nimrod in the flesh, his admonition would be, "Keep to leeward and stalk carefully, breaking no twig, for your game will run away if it winds or hears you." That is the first rule I learned for big-game hunting, and it is founded on fact. But there are two other facts I have observed these many years, which Nimrod will never mention: the first, that when you are keenly hunting, it often happens that game breaks away in alarm *before* it winds or sees or hears you; and the second, that when you are not hunting, but peaceably roving the woods, going carelessly and paying no attention to the wind, you often come very close to wild game, which stops to watch you curiously *after* it has seen you and heard your step or voice and sampled your quality in the air. These two facts, implying some kind of mental or emotional contact between the natural man and the natural brute, are probably not accidental or unrelated, and we are here trying to find the natural law or principle of which they are the occasional and imperfect expression.

This whole matter of silent communication may appear less strange if we remember that most wild creatures are all their lives accustomed to matters which sense-blinded mortals find mysterious or incredible. Why a caterpillar, which lives but a few hours when all the leaves are green, should make a cocoon of a single leaf and with a thread of silk bind that leaf to its stem before he wraps himself up in it, as if he knew that every leaf must fall; or why a spider, adrift for the first time on a chip, should immediately send out filaments on the air currents and, when one of his filaments cleaves to something solid across the water, pull himself and his raft ashore by it; or why a young bear, which has never seen a winter, should at the proper time prepare a den for his long winter sleep,—a thousand such matters, which are as A B C to natural creatures, are to us as incomprehensible as hieroglyphics to an Eskimo. That a sensitive animal should know by feeling (that is, by the reception of a silent message) whether an approaching animal is in a dangerous or a harmless mood is really no more remarkable than that he should know, as he surely does, when it is time for him to migrate or to make ready his winter quarters.

This amazing sensitiveness, resulting, I think, from the reception of a wordless message, was brought strongly home to me one day as I watched a flock of black mallards, forty or fifty of them, resting in the water-grass within a few yards of my hiding-place. A large hawk had appeared at intervals, circling over the marshes and occasionally over the pond; but, beyond turning an

eye upward when he came too near, the ducks apparently paid
no attention to him. He was their natural enemy; they had paid
toll of their number to satisfy his hunger; but now, though
plainly seen, he was no more regarded or feared than a dragon-
fly buzzing among the reeds. Presently another hawk appeared
in the distance, circling above the meadows. As a wider swing
brought him over the pond a watchful duck uttered a single low
quock! On the instant heads came from under wings; a few
ducks shot into the open water for a look; others sprang aloft
without looking, and the whole flock was away in a twinkling. I
think the hawk did not see or suspect them till they rose in the
air, for at the sudden commotion he swooped, checked himself
when he saw that he was too late, and climbed upward where he
could view the whole marsh again.

Now these two hawks were of the same species, and to my
eyes they were acting very much alike. One was hungry, on the
lookout for food; the other was circling for his own amusement
after having fed; and though the eyes of birds are untrustworthy
in matters of such fine distinction, in some way these ducks
instantly knew or felt the difference between the mood of one
enemy and that of another. Likewise, when I have been watching deer in winter, I have seen a doe throw up her head, cry an
alarm and bound away; and her action became comprehensible
a few moments later when a pack of hunting wolves broke out
of the cover. But I have watched deer when a pack of wolves
that were not hunting passed by in plain sight, and beyond an
occasional lift of the head for an alert glance the timid creatures
showed no sign of alarm, or even of uneasiness, in presence of
their terrible enemies.

I say confidently that one wolf pack was hunting and the other
not hunting because the northern timber-wolf naturally (that is,
in a wild state and dealing with wild animals) hunts or kills only
when he is hungry. I ran the trails of both packs, and one
showed plainly that the wolves were in search of food; while the
other said that the brutes were roaming the country idly, lazily;
and when I ran the back trail of this second pack I found where
they had just killed and eaten. The deer were not afraid of them
because they were for the time quite harmless.

At first I thought that these ducks and deer perceived the mood of their enemies in a simple way through the senses; that they could infer from the flight of a hawk or the trot of a wolf whether he were peaceable or dangerous; and at times this is probably the true explanation of the matter. The eyes of most birds and beasts, strangely dull to objects at rest, are instantly attracted to any unusual motion. If the motion be quiet, steady, rhythmical, they soon lose interest in it, unless it be accompanied by a display of bright color; but if the motion be erratic, or if it appear and disappear, as when an approaching animal hides or creeps, they keep sharp watch until they know what the motion means or until timidity prompts them to run away. Thus, chickens or ducks show alarm when a kite slants up into the air; they lose interest when the kite sits in the wind, and become alert again when it begins to dive or swoop. It is noticeable, also, that on a windy day all game-birds and animals are uncommonly wild and difficult of approach, partly because the constant motion of leaves or grass upsets them, and partly (in the case of animals) because their noses are at fault, the air messages being constantly broken up and confused. But such a "sensible" explanation, suitable as it may be for times or places, no longer satisfies me, and simply because it does not explain why on a quiet day game should be unconcerned in presence of one hawk or wolf, and take to instant flight on the appearance of another enemy of the same species.

It should be noted here that these "fierce" birds and beasts are no more savage in killing grouse or deer than the grouse is savage in eating bugs, or the deer in seeking mushrooms at the proper season; that they simply seek their natural meat when they are hungry, and that they are not bloodthirsty or ferocious or wanton killers. Only men, and dogs trained or spoiled by men, are open to that charge. The birds and beasts of prey when not hungry (which is a large part of the time, since they feed but once a day or sometimes at longer intervals) live as peaceably as one could wish. After feeding they instinctively seek to be with their own kind and very rarely attempt to molest other creatures. At such times, when they are resting or playing or roving the woods, the smaller wood folk pay no more attention to them than to harmless fish-hawks or porcupines.[4] Repeatedly I have watched game-birds or animals when their enemies were in sight, and have wondered at their fearlessness. The interesting question is, How do they know, as they seemingly do, when the full-fed satisfaction of their enemy changes to a dangerous mood? Why, for example, are deer alarmed at the yelp of a she-wolf calling her cubs to the trail, and why do they feed confidently in the dusk-filled woods, as I have seen them do, when the air shivers and creeps to the clamor of a wolf pack baying like unleashed hounds in wild jubilation?

I have no answer to the question, and no explanation except the one suggested by human experience: that the hunting animal, like the hunting man, probably sends something of his emotional excitement in a wave ahead of him, and that some animals are finely sensitive enough to receive this message and to be vaguely alarmed by it.

The mating of animals, especially the calling of an unseen mate from a great distance, brings us face to face with the same problem, and perhaps also the same answer. Sometimes the mating call is addressed to the outer ear, as in the drumming of a cock-grouse or the whine of a cow-moose; but frequently a mate appears when, so far as we can hear, there is no audible cry to call him. How do the butterflies, for example, know when or where to seek their other halves? That their meeting is by chance or blunder or accident is a theory which hardly endures an hour's observation. In the early spring I take a cocoon from a certain corner of shrubbery and carry it to my house, and there keep it till the end softens, when I put it into a box with a

---

[4] In parts of the West, I am told, wolves often kill more than they need. Formerly they fed on the abundant game and were wholly natural animals; but their habits have changed with a changed environment. When the game was destroyed by settlers or hunters the wolves began to feed on domestic animals; and the descendants of these wolves, which killed right and left in a crowding, excited herd of sheep or cattle, are now said to kill deer wantonly when they have the chance. I cannot personally verify the saying, and know not whether it rests on exceptional or typical observation. In the North, where there are no domestic animals, I have rarely known a timber-wolf to kill after his hunger was satisfied.

screened top and hang it out under the trees. Presently a gorgeous moth crawls out of the cocoon; and hardly has she begun to wave her wings to dry them when the air over the screen is brilliant with dancing wings, the wings of her would-be mates. And the thing is more puzzling to me because I have never found a cocoon of that kind in my immediate neighborhood; nor have I seen a single *cecropia* this season until the captive called them.

How they find her so promptly is a problem that I cannot solve. It may be that the call is wholly physical or sensible, that some fine dust or aroma is sent forth on the air currents, and the sensitive nerves of other moths receive and respond to it; but it is still amazing that wind-blown creatures can follow an invisible air-trail through what must be to them a constant tempest and whirlwind of air currents, until they come unerringly to the one desired spot in a limitless universe. I have shown that pretty sight of dancing wings to many audiences, after predicting what would happen; and always they saw it with wonder, as if there were magic in it.

The moth mystery may be dissolved by some such purely physical formula; but what physical sense will explain the fact that when I turned a modest hen-pheasant loose in the spring, in a region where my wide-ranging setter and I never discovered a pheasant, she was immediately joined by a gloriously colored mate, and soon there was a hidden nest and then young pheasants to watch? Most birds and beasts are questing widely in the mating season, and their senses seem to be more keen at this time, or more concentrated on a single object. On grounds of what we thoughtlessly call chance, therefore, they would be more apt to find mates when they are keenly looking for them; but giving them every possible chance in a wide region where the species is almost extinct, and then multiplying that chance a hundred times, I still find it hard to believe that the meeting of two rare animals is either accidental or the result of ordinary sense-perception. Out of several examples that occur to me, here are two which especially challenge the attention:

One early spring a she-fox was caught in her den, some five miles from the village where I then harbored. She was carefully bagged, carried a few rods to an old wood road, placed in a wagon and driven over country highways to the village, where she was confined in a roomy pen in a man's dooryard. A few nights later came a snowfall, and in the morning there were the tracks of a male fox heading straight to the vixen and making a path round about her pen. She was his mate, presumably, and when we found his tracks our first feeling of admiration at his boldness was soon replaced by the puzzling question of how he had found her so quickly and so surely. To answer that question, if possible, I followed his back trail.

Now the trail of a fox in the wilderness, where he is some-times hunted by wolves or other hungry prowlers, is a bewilder-ing succession of twistings and crisscrosses; in a settled region, where his natural enemies are extinct, his trail is bolder, more straightforward, easier to read; and in either case you can quick-ly tell by the "signs" whether your fox is male or female, wheth-er hunting or roaming, or hungry or satisfied. Also you can tell whether he is just "projeckin' around," as Uncle Remus says, or whether his mind is set on going somewhere. In the latter event he almost invariably follows runways, or fox roads, which are as well known to him as are footpaths and stream-crossings to a country lad. But the trail of this particular fox was different from any other that I ever followed. That he was a male and was "going somewhere" was evident enough; but he was not follow-ing runways or paying any attention to them. He left no signs at places where any ordinary dog-fox would surely have left them, and he was stopping to listen or to ward himself at uncommonly frequent intervals. So, running it backward, I read the story of his journey mile after mile, till the oncoming trail changed to the devious, rambling trot of a questing fox; and beyond that I had no interest in it.

The place where the fox seemed to have found his bearings, or where he stopped his rambling to head straight for his mate, was some four miles distant from the captive in a bee-line. The course he took was entirely different from that taken by the man who brought the vixen home, thus excluding the theory that he followed the trail by scent; and the latter part of his way led through the outskirts of a village, where the track of a fox had not been seen for many years. From the distant hills he had come down through sheltering woods at a stealthy trot; across open pastures on the jump; over a bridge and along a highway, where he traveled behind a friendly stone wall; then very cautiously through lanes and garden fringes, where the scent of men and dogs met him at every turn; turning aside here for a difficulty or there for a danger, but holding his direction as true as if he followed a compass, till he came at last with delicate steps to where his mate was silently calling him. For except on the assumption that she called him, and with a cry that was soundless, I know not how to explain the fact that he found her in a place where neither he nor she had ever been before.

It is possible, you may reason, that this was not his first visit; that unknown to us, venturing among his human and canine enemies, he had by a lucky chance stumbled upon his mate on an evening when the bare ground did not betray his secret to our eyes; and that for his next visit he had cunningly laid out a different trail through manifold dangers. It was the latter trail, made without doubt or question of what lay at the end of it, which I had followed in the telltale snow.

That is a good armchair argument, but a very doubtful explanation of the fox action, since it calls for more reasoning power than we commonly find in a brute. Remember that this male fox was far away on his own affairs when his mate was captured, and he had no means of knowing where she had gone: he simply missed her from her accustomed haunts, and sought till he found her. Remember also that a male fox is never allowed to come near his mate's den; that when she is heavy with young, as this vixen was when caught, he may join her of an evening and hunt for her, or bring her food that he has killed, but always, I think, at a distance from the place where she intends to bring forth her cubs. That the male, after missing his mate and yapping for her in vain through the woods, should at last seek her at the forbidden den, and there find the scent of men and conclude that they had taken her away; that he should follow the scent of a wagon-wheel over five miles of country roads, or else explore all the neighboring villages till he found what he sought; that he should then lay out a different trail, more secret and more reasonably safe, for his second visit,—all this is to me more ingenious, more unnatural and more incredible than that his mate should silently call him, as his mother had probably many times called him from the den when he was a cub, and that he should feel and answer her summons.

Such reasoning is purely speculative, but there are certain facts which we must keep in mind if we are to explain the matter. The first, a general fact which is open to observation, is that it is fox nature at certain seasons to come to a captive; for what reason or with what self-forgetful motive it would be hard to say. I have known mother foxes and mother wolves to come where their cubs were imprisoned by men. I have heard a straight record of one male wolf that appeared at a ranch the second night after his wounded mate was captured by the ranchman. And I have seen a male fox come to the rescue of a female when she was driven by dogs and too heavy with young to make a long run, and wait beside her trail till the dogs appeared, and then lead them off after him while she made her escape. The second fact, which may imply some power of silent communication, is that when snow fell about the pen of this captive fox, a few nights after she was taken, there were the tracks to show that her mate already knew where she was. That he found and came to her in the midst of his enemies may be quite as significant as *how* he found her, by way of giving a new direction or interest to our skin-and-bones study of natural history.

In this first example the fox was perhaps moved by the mating impulse, which sharpened his wit and encouraged his will; but at times a wild creature may seek and find his mate when no such "call of the blood" urges him, when he comes with his life in his hand in response to some finer or nobler motive, something perhaps akin to loyalty, which is the sum of all virtues. Witness the following:

A friend of mine was hunting one October day when his shot wing-tipped a quail, apparently the old female of the flock, which his dog caught and brought to him almost uninjured. What to do was the next question. It is easy, because thoughtless, to cut down a bird in swift flight; but when the little thing nestles down in your hand or tries to hide under your fingers;

when you can feel its rapid heart-beat, and its eyes are big and wondrous bright,—well, then some hunters bite the head, and some wring the neck, and some would for the moment as lief be shot as to do either. So to avoid the difficulty my friend put the quail carefully away in a pocket of his hunting-coat, and brought her home with some vague idea of taming her, and some dream of trapping a mate in the spring and perhaps raising some little bob-whites of his own. At night he put the captive into a coop just inside the barn window, which was open wide enough to admit air but not a prowling cat; for he was already beginning to learn that a quail is a most lovable little pet, and he was bound that she should not again be hurt or frightened.

Before sunrise next morning he heard low, eager whistlings in the yard; and there was another quail, a male bob-white, where never one was seen before nor since. He was perched warily on the window-sill of the barn, looking in at his captive mate, telling her in the softest of quail tones (for there were enemies all about) that he had found her and was glad; while from within the barn came a soft piping and gurgling which seemed to speak welcome and reassurance. The opening of a door frightened him; he buzzed away to the orchard, and presently from an apple-tree came the exquisite *quoi-lee! quoi-lee!* the assembly-call of the bob-white family.

The first quail had been caught miles away from the man's house; there were no other birds of her own kind within hearing distance, so far as my friend and his dog ever discovered, and it was not the mating-time, when quail are questing widely. By a process of elimination, therefore, one reaches the conclusion that the male bird was, in all probability, the father of the flock over which the captive presided; that he had helped to raise the young birds, as quail commonly do; that he had stood by his family all summer in loyal bob-white fashion, and that he went out to seek his vanished mate when she failed to answer his calling.

All that seems clear enough, and in perfect accord with quail habits; but when you ask, "How did he find her?" there is no answer except the meaningless word "chance" or a frank admission of our original premise: that wild birds and beasts all exercise a measure of that mysterious telepathic power which reappears now and then in some sensitive man or woman. It may be for the wood folk that "There is no speech nor language: their voice is not heard," as the Psalmist wrote of the communing day and night; but they certainly communicate in some way, and the longer one studies them the more does it appear that part of their "talk" is of a finer character than that which our ears can hear.

VII

To know birds and beasts may be a greater or a lesser triumph than to know ornithology or zoölogy. That is a question of taste or temperament, the only certainty being that the two classes of subjects are altogether different. The latter deals with external matters, with form, classification, generalities. Its materials are books, specimens, museums, one as dead or desiccate as another; and because it is limited and exact, you can memorize its outlines in a few days, or become in a few years an authority in the science.

The former subject, of birds and beasts, deals with an endless and fascinating mystery. Its materials are living and joyous individuals, among whom are no classes or species, concerning whom there can be no "authority"; and when, after a lifetime of study, you have made a small beginning of knowledge, you find that, like All Gaul of misty memory, it may be divided into three parts.

One part is observation, which is a simple matter of the eye. Another is sympathy, which belongs to the mind or heart. In dealing with wild creatures, as with civilized folk, one learns to appreciate De Quincey's rule of criticism "Not to sympathize is not to understand." A third part, more rare and variable, may come from that penetrating but indescribable quality which we call a gift. A few men have it; the animals instinctively trust them, and they understand animals without knowing how they understand. The rest, lacking it, must struggle against a handicap to learn, substituting the slow wisdom of experience for the quick insight of the gift. It is for the latter chiefly that I write these wood notes.

One word more by way of preface, to express the conviction that you can learn nothing worth knowing about birds or beasts so long as you seek them with a gun in your hand. On that road you shall find only common dust, and at the end of it a valley of dry bones. Whether you carry the gun frankly for sport, or delude yourself with the notion that you can add to natural history by collecting more skins or skulls, you have unconsciously placed destruction above fulfilment, stark death above the beautiful mystery of life. So must you estrange both the animal and yourself, making it impossible for you to meet on any common ground of understanding. And now for our lessons:

If I were to formulate rules for a subject which can never be learned by the book, I might say that there are three things you should know, and another you must do, if you expect to gain any intimate knowledge of the wood folk, or even to approach them near enough for fair and leisurely observation.

The first thing to know is that natural creatures, though instinctively shy or timid, are not wildly governed by fears and terrors, as we have been misinformed from our youth up. The "reign of terror" is another of those pet scientific delusions, like the "struggle for existence," for which there is no basis in nature. Fear in any true sense of the word is an exclusively human possession, or affliction; it is a physical and moral poison, as artificial as sin, which the animal escapes by virtue of being natural. It is doubtful, indeed, whether anything remotely resembling our fear, a state of mind arising from a highly developed imagination which enables us to picture events before they happen, is ever born into a hairy skin or hatched out of an egg. The natural timidity of all wild creatures is a protective and wholesome instinct, radically different from the fear which makes cowards of men who have learned to trace causes and to anticipate consequences.

So much for the mental analysis; and your eyes emphasize the same conclusion when you look frankly upon the natural world. The very attitude or visible expression of birds or beasts when you meet them in their native woods, feeding, playing, resting, seeking their mates, or roving freely with their little ones (all pleasurable matters, constituting nine-tenths or more of animal existence), is enough in itself to refute the absurd notion of a general reign of terror in nature. If you are wise, therefore, you will get rid of that prejudice, or at least hold it in abeyance till the animals themselves teach you how senseless it is. To go out obsessed with the notion of fear is to blind your eyes to the great comedy of the woods.

The second thing to know, and to remember when you go forth to see, is that sensitive creatures dislike to be watched, and become uneasy when they find a pair of eyes intently fixed upon them. You yourself retain something of this ancient animal inheritance, it seems, since there is nothing which more surely excites alarm if you are timid, or challenge if you are well balanced, or anger if you have a fighting spirit, than to have a stranger watching your every move while you go about your lawful affairs. The fact that you cannot word a reason for your alarm or challenge or anger makes you all the more certain that you have an unanswerable reason; which is your inborn right to be let alone.

This natural and inalienable right (which society curbs for its own protection, and reform societies trample on for their peculiar pleasure) may help you to understand why the animal becomes alarmed when he finds you watching him closely. He desires above all things else, above his dinner even, to be let alone; and your eye may as surely disturb his peace, his self-possession, his sense of security, as any gun you may shoot at him or any fire you may kindle in his fragrant domain.

You have but to think a moment in order to understand why even your look may be too disturbing. When a beast of prey sees a buck that he wants to catch, what is his invariable mode of procedure? First he hides, then he creeps or skulks or waits, all the while keeping his eyes fastened upon his victim, watching every move with fierce intensity till the moment comes to spring. It follows, naturally enough, that when the same beast of prey finds other eyes fixed upon himself, he knows well what the look means, that a rush will swiftly follow; and he anticipates the rush by taking to his heels. Or the buck, having once escaped the charge of a hunting beast, will remember his experience the next time he finds himself an object of scrutiny, and will flee from it as from any other discomfort.

Whether this action is the result of instinctive or deductive knowledge is here of no consequence: let the psychologists pick a bone over it. Since we have in our heads a strong aversion to being observed too closely, we are probably facing an instinct, which is stronger in the brute than in the man; but it is the fact, not the explanation thereof, which is important. The simple fact is, that wild birds and beasts will not endure watching; and you begin to sympathize with their notion when you mark the eyes of a stalking cat, with their terrible fire just before she springs. There is always more or less of that fire in a watchful eye; you may see it glow or blaze under a man's narrowed lids before he takes quick action; and it is the kindling of that dangerous light which a sensitive creature expects and avoids when he finds you watching him.

Did you ever follow an old cock-partridge in the woods with intent to kill him? If so, you have a living picture of the truth I have explained theoretically. Near our towns the partridge (ruffed grouse) is very wild, taking wing at your approach; but in the deep woods he is almost fearless. Even when you stumble into a flock of the birds, frightening them out of their calm, they are apt to flit into the trees and remain absolutely motionless. They are then hard to find, so well do they blend with their background; and if they are young birds, they will hold still after you discover them. Since they were helpless chicks they have trusted to quietness to conceal them; it serves them very well, much better than running away from stronger enemies; and the habit is strong upon them, as upon young ducks and other game-birds before they have learned to trust their wings. But when you stumble upon an old cock-grouse you meet a bird that has added experience to instinct, and that knows when to move as well as when to sit still. He dodges out of sight as you raise your rifle; as you follow him he bursts away on whirring wings and slants up into a tree in a distant part of the wood. Marking where he lights, you try to find him, cat-footing around his perch, peering into every tree-top, putting a "crik" in the back of your neck. For a half-hour, it may be, you search for him in vain; suddenly there he is, and—*b'r-r-r-r!* he is gone. The odd thing is that he sits still so long as you cannot find him; not a feather stirs or a foot shifts or an eyelid blinks even when your glance roves blindly over him; you may give him up and go away, leaving him motionless; but the instant you see him he seems to know it, and in that instant he is off. This is not a single or an accidental but a typical experience; any woodsman who has hunted ruffed grouse with a rifle will smile as he tells you, "That's true; but I can't explain it."

A third bit of woods lore, of which we shall presently make good use, is that natural birds and animals have a lively interest in every new or strange thing they meet. Far from being occupied in a constant struggle for existence, as the books misinform us, their lives are full of leisure; they have plentiful hours for rest or play or roving, and in these idle times they get most of their fun out of life by indulging their curiosity. I fancy that in this respect, also, most people are still natural creatures, seeing that men or women in a crowd are as easily set to stretching their necks as any flock of ducks or band of caribou.

So strong is the animal's inquisitive instinct (for it surely is an instinct, the basis of all education, and without it we should be fools, learning nothing) that he will readily give over his play or even his feeding to investigate any new thing which catches his attention. I speak now not of fearsome things, which may properly alarm the wood folk, but of pretty or harmless or attractive things, such as the repeated flash of a looking-glass or the rhythmic swing of a handkerchief or a whistled tune, which commonly bring wild creatures nearer with forward-set ears and eyes with questions in them. In a word, so far as I have observed birds and beasts, their first or natural attitude toward every new object, unless it be raising fearful smells or moving toward them with hostile intent, is invariably one of curiosity rather than of fear.

One proof of this universal trait, to me, is that when I approach wild animals carelessly they often run away; but of the hundreds that have approached me when I was quiet in the woods, every one without exception showed plainly by his action that he was keen to find out who or what I might be. Young animals are more inquisitive than old, having everything to learn, and they are easily attracted; but age cannot stale the wonder of the world for them, and I have never chanced to meet an old doe, no, nor a tough old bull moose, that did not come near to question me if the chance were given. Of the larger wood folk Mooween the bear is perhaps the least inquisitive; yet once an old bear came so close to me, his eyes a question and his nose an exclamation point, that I could have touched him before his curiosity was satisfied; and several times, when I have been watching the berry-fields, a bear and her cubs have noticed some slight motion of mine and have left their feast of blueberries to approach rather too near for my comfort. At close quarters an old she-bear is a little uncertain. Commonly she runs away in sudden panic; but should you get between her and her cub, and the piggish little fellow squeal out as if frightened or hurt, she may fly into a fury and become dangerous to a man unarmed.

The obvious thing to do, in view of what has been learned, is to hold physically and mentally still when you meet a wild animal, and so take advantage of his curiosity. That is very easy when he happens to find you at rest, for then he is bound to find out something about you before he goes; but even when he catches you afoot you may still have a fair chance if you stop in your tracks and move no muscle while he is looking. Remember that so long as you are motionless you puzzle him; that you should advance only when his head is turned away, and that you should never move directly at any animal, but to one side, as if you would give him plenty of room in passing. If you must change your position or attitude while he is looking, move gently and very slowly, avoiding every appearance of haste or nervousness. If he vanishes after one keen look, be sure he is a veteran that has seen men before, and bide still where you are. The chances are ten to one that no sooner does he think himself hidden than he will turn to have another look at you. It is always in your favor, since you have the better eyes, that an animal has the habit of concealment, and so long as you pretend not to see him he is very apt to think himself unseen.

Such a method applies particularly well to all members of the deer family, with their insatiable curiosity; but it serves almost as well with beasts of prey, which may be so surprised by meeting a motionless man that they will often "point" him in a way to suggest a setter pointing a woodcock. We think of the fox, for example, as the most cunning of animals; like the dolls' dressmaker in *Our Mutual Friend,* he seems to be saying, "Oh, I know your tricks and your manners"; yet on a good tracking-snow I have trailed many foxes to their day-beds, and have found that with few exceptions they act in the same half-puzzled, half-inquisitive way. And this is the fashion of it:

Looking far ahead on the dainty trail you suddenly catch a glimpse of orange color, very warm against the cold whiteness of the snow, which tells you where Eleemos the sly one, as Simmo calls him, is curled on a warm rock or stump with the winter sunshine fair upon him. Then you must leave the trail, as if you were not following it, and advance on noiseless feet till the fox raises his head, when you must "freeze" in your tracks. If he is a tramp fox (that is, one which has come hunting here out of his own territory) or a veteran that has already seen too much of men and their devices, he will dodge out of sight and be seen no more; but if he is an ordinary young fox, especially a cub weathering his first winter, he will almost certainly investigate that odd motionless object which was not there when he went to sleep. After "pointing" you a moment he slips into the nearest cover, not turning his head in your direction, but watching you keenly out of the corners of his yellow eyes. When he thinks himself hidden from your sight he circles to get your wind; and on this side or that you will have two or three good glimpses of him before he floats away—or seems to, so lightly does he run—

to hunt up another day-bed. Your last view of him shows a slyly inquisitive little beast, perfectly self-possessed; but as he disappears you notice a nervous, quivering, fluttering motion of his great brush, which gives him away as a tail betrays a dog, and which says that Eleemos is greatly excited or puzzled over something.

Better than roaming noisily through the woods in search of game is to sit still and let the game come to you—an arrangement which puts you at your ease, and at the same time encourages the animal to indulge his curiosity without alarm. You may not see so many birds or beasts in this way, but some of them you shall see much more intimately; and a single inquisitive jay may teach you more of nature than all the bird books in the world, as I have learned more of Latin humanity from Angelo, who polishes my shoes, than from Gibbon's *Decline and Fall of the Roman Empire.* Very often, if you hold perfectly still, a wild animal will pass down the runway close at hand without even seeing you, and you must draw his attention by a chirp or a slight motion. Then, when he whirls upon you in astonishment, his eyes saying that he was never so surprised in his life, observe him casually as it were, veiling your interest and never staring at him as if he were a wild or strange beast, but greeting him rather as one you have long known.

At such a moment quietness is the best medicine—quietness and friendly eyes. If the animal wavers, a low song or a whistled tune may or may not be helpful; it depends entirely on the tune. You are to keep physically quiet, because any sudden motion will alarm the sensitive creature, so near is he to the unknown; and mentally quiet, because excitement is as contagious as fear or measles, or any other disease of mind or body. When I am alone in the woods wild animals are rarely hard to approach, and when I am sitting quietly by a runway they show no fear of me whatever, drawing near with questioning eyes or moving away reluctantly; but when I take another with me, especially one who grows excited in the presence of big game, the same animals appear suspicious, uneasy, and end by bolting away as if we had frightened them.

One day there came to my camp a friend who was eager to

see a deer at close range, but who was doubtful of my assurance that animals could neither see nor smell him if he knew how to hold still. When I promised him a deer at ten feet he jumped for his camera, saying that in such an incredible event he would get what he had always wanted, a picture of the graceful creature against a background of his native woods, in soft light and shadow instead of the glaring black-and-white of a flashlight. At that disturbing proposition all his doubts moved into me, who have always found camera folk a fidgety folk. What with their fussing and focusing and everlasting uneasiness over distance or time or shutter, or something else which is never right and ready, they are sure to bedevil any wild creature before he comes within speaking distance; so I took my friend and his camera along without faith, hoping for the best.

Our stand was a hardwood ridge where deer often passed on their way to the lake, and we had been sitting there hardly an hour when I saw a young spikebuck coming down the runway. The next moment there was a gasping "Oh, there's a deer!" from the man who had been warned to keep mentally still. Then began the inevitable tinkering with the camera, which had been thrice prepared and was still as unready as all its kind. More than once I had sat in that precise spot while deer passed at a distance of three or four yards without noticing me; but now the little buck caught an uneasy motion and halted with head high and eyes flashing. If ever there was a chance for a wonderful picture, he offered it; but he did not like the focusing, or whatever it was, and after endless delay the camera clicked on a white flag bobbing among the shadows, where it looked in the negative like a smear of sunlight.

The camera reminds me of another way of approaching deer, a way often followed by summer campers; namely, by chasing the swimming animal in a canoe. I have but one word to say of such a method, and that is, Don't! When a deer is crossing broad water you can get as close to him as you will; you can take a grip on him and let him tow your canoe, as thoughtless people sometimes do, encouraged by their guides; but I suggest that it would be much better to shoot the creature and have done with it.

A deer's powers are very delicately balanced; he is nervous, high strung, easily upset. Even on land, where he can distance you in a moment, he begins to worry if he finds you holding steadily to his trail; and I have known a young deer to become so flustered after he had been jumped and followed a few times that he began to act in most erratic fashion, and was very easily approached. When you chase him in the water, and he finds that he cannot get away from you, he may give up and drown, as a rabbit submits without a struggle when a weasel rises in front of him; but a vigorous deer is more apt to become highly excited, to struggle wildly, to waste ten times as much energy as would keep him afloat, to jump his heart action at a dangerous rate; and then a very little more will finish him as surely as a bullet in the brain.

Twice have I seen deer thus killed by thoughtless campers, the last victim being a splendid buck full grown. Two men saw him swimming an arm of Moosehead Lake, and launched a canoe with no unkinder purpose than to turn him back to shore, so that a party of sportsmen there might get a picture of him. The buck labored mightily; but the paddles were swift, and wherever he turned the danger appeared close in front of him. Suddenly he rose in the water, pawing the air, and heaved over on his side. When the canoe reached him he was dead; and the surprising thing is that dissection revealed no ruptured bloodvessel nor any other visible cause of his death. It was probably a matter of heart paralysis. Such an ending was unusual, I know; but undoubtedly many of these overwrought animals reach shore exhausted, spent to the limit, and lie down in the first good cover, never to rise again.

Moose and caribou are stronger swimmers than deer, and of tougher fiber; but it is still dangerous, I think, to chase them in the water. Once I saw a canoe following close behind a cow and a calf moose, the canoeists yelling wildly to hurry up the pace. Had they thought to look once into the eyes of the struggling brutes, they might have learned something which they ought to know. As the calf lagged farther and farther behind, the mother turned to come between him and the canoe, and remained there trying to urge and push the little fellow along. So they reached shallow water at last, found their footing, and plunged into the cover. The canoe turned away, and no doubt the incident was soon forgotten. I never saw the canoemen again, but I saw one of the moose. A few days later, in passing through the woods on that side of the lake, I found the calf stretched out dead where he had fallen, not fifty yards from the water's edge.

Perhaps another "don't" should here be memorized for the happy occasion when you find a fawn or a little cub in the woods, and are moved most kindly to pet him. If the mother is half-tame, or has lived near a clearing long enough to lose distrust of the man-scent, it may do no harm to treat her fawn or cub as you would a puppy; but to handle any wild little creature is to do him an injury. Until a fawn is strong enough to travel the rough country in which he was born, the doe often leaves him hidden in the woods, where he lies so close and still that you may pass without seeing him. Once you discover him, however, and he knows that he is seen, his beautiful eyes begin to question you with a great wonder. He has no fear of you whatever (this while he is very young, or before he begins to follow his mother); he will sometimes follow you when you go away, and he is such a lovable creature, so innocent and so appealing, that it is hard to keep your hands from him. Let him sniff your palm if he will, or lick it with his rough tongue for the faint taste of salt; but as you value his life don't pet him or leave the scent of you on his delicate skin. A wild mother knows her own by the sense of smell chiefly; if she finds the startling man-scent where she expected a familiar odor, she becomes instantly alarmed, and then the little one is a stranger to her or a source of violent anger.

Once, before I learned better than to handle any helpless cub, I saw a doe drive her own fawn roughly away from her, out of my sight and hearing. I had petted the fawn a little (a very little, I am glad to remember) and looked with wonder on the mother's anger, not understanding it till some time later, when I learned of a similar incident with a sadder ending. Not far from my camp a sportsman with his guide found a fawn hidden near the stream where they were fishing, and being completely won by the beautiful innocent, as most men are, they petted him to their hearts' content. When an old doe, the mother presumably, appeared heading in their direction they thoughtfully withdrew, hiding at a distance to watch the family reunion. The doe seemed to hasten her steps when she saw that the fawn was on his feet, instead of lying close where she had left him; but when near him she suddenly stiffened, with the hair bristling on her neck. Two or three times she thrust out her nose, only to back away, and once she raised the harsh alarm-cry that a doe utters when she smells danger. Then, as the little fellow trotted up to her on his wabbly legs, she leaped upon him in fury and trampled him to death.

On Keeping Still

VIII

To return to our first lesson, of quietude: it was impressed upon me unconsciously, like most good lessons, before I had any thought that I was learning the true way of the woods. The teacher was Nature herself (she seldom fails to quiet boy or man if left alone with him), and the school-room was a lonely berry-pasture surrounded by pine and hardwood forests. The berry-pickers, a happy and carefree lot, often let me go with them while I was yet too small to find my way among the tall swamp-blueberry bushes, and would leave me under a tree at the edge of the woods, with an armful of berry-laden branches to keep me busy while they wandered far away in search of the best picking. Sitting there in the breathing solitude, occupied with the task of filling my tin cup with berries and well content with my lot (for the woods always had a fascination for me, and seemed most friendly when I was alone), I would presently "feel" that something was watching me. There was never any suggestion of fear in the impression, only an awakening to the fact that I was not alone, that some living thing was near me.

119

Then, as I looked up expectantly, I would almost always find a bird slipping noiselessly through the branches overhead, or a beastie creeping through the cover at my side; and in his bright eyes, his shy approach, his withdrawal to appear in another spot, I read plainly enough that he was asking who I was or what I was doing there. And by a whistled tune or a drumming on my cup, or by flashing a sunbeam into his eyes from a pocket glass, I always tried to hold him as long as I could.

This curious sense or feeling of being watched, by the way, is very real in some men, who do not regard it as a matter of chance or imagination. I have known of two elaborate courses of "laboratory" experiments which aimed to determine how far such a feeling is trustworthy, and both resulted in a neutral or fifty-fifty conclusion; but I wonder, if the experiments had been tried on Indians or natural men under natural conditions, whether the result might not have been quite different. The fact that the first fifty men you meet get lost or turned around in a trackless forest is significant for the fifty, and for the vast majority of others; but it means nothing to the one bushman who can go where he will without thought or possibility of being lost, because of his sure sense of direction.

So, possibly, with this feeling of being watched: it may be too intangible for experiment, or even for definition. Many times since childhood when I have been alone in the big woods, fishing or holding vigil by a wilderness lake, I have the feeling, at times vaguely and again definitely, that strange eyes were upon me. Occasionally, it is true, I have found nothing on looking around, either because no animal was there or because he was too well hidden to be seen; but much more often the feeling proved true to fact—so often, indeed, that I soon came to trust it without doubt or question, as Simmo my Indian still does, and a few other woodsmen I have known. It is possible that one's ears or nose may account for the feeling; that some faint sound or odor may make itself felt so faintly that one has the impression of life without knowing through what channel the impression is received. Of that I am not at all sure; at the moment it seems that some extra sense is at work, more subtle than smell or hearing; and, whether rightly or wrongly, it is apparently associated with the penetrating stare of an animal's eyes on your back.

To quote but a single incident, out of several that come to my memory: I was once sitting on the shore of a lake at twilight, wholly intent on following the antics of a bull moose I had called into the open. He was on the other side of a small bay, ranging up and down, listening, threshing the bushes with his antlers, blowing his penny-trumpet at intervals,—in a dozen impatient ways showing what a young and foolish moose he was. A veteran would have kept to the cover till he had located what he came for. I had ceased my bellowing when the bull first answered, had been thrilled by his rush through the woods, had cheered him silently when he burst into the open, grunting and challenging like a champion; now I was quietly enjoying his bewilderment at not finding the tantalizing cow he had just heard calling. He did not see or suspect me; I had the comedy all to myself, and was keenly interested to know how he would act when he rounded the bay, as he certainly would, and found me sitting in his path. Because he was big and truculent and a fool, I did not know what to expect; my canoe floated ready against the outer end of a stranded log, where a push would send it and me into deep water. I mention these details simply to show where my thoughts were.

As I watched the play in the hushed twilight, suddenly came the feeling that something was watching me. The bull had started around the bay in my direction; possibly his eyes had picked me out—but no, he was in plain sight, and the feeling is always associated with something unseen. Without changing position I looked carefully all about, searching the lake and especially the woods, which were already in deep shadow. Finding no bird or beast, no motion, nothing alarming, I turned to question the bull, who had halted to sound his ridiculous trumpet. He was perhaps fifty or sixty yards away. He had not yet seen me; I had no fear of him, no anxiety whatever; yet again came the feeling, this time insistent, compelling, as if some one had touched me and said, "*Get away!*" I did so promptly, jumping to my feet; and out of a fir thicket behind me charged another bull that I had not dreamed of calling.

By his size, his antlers, his fierce grunting, I recognized this brute on the instant. I had met him before, once on a trail, once on the lake shore, and had given him all the room he wanted. He was a grizzled old bull, morose and ugly, that seemed to have lost his native fear of man—from a galling wound, perhaps, or from living an outcast life by himself. He was a little crazy, I judged. That he was dangerous I knew from the fact that he had previously made an unprovoked attack upon my Indian. He, too, had heard the call; had approached it from behind as stealthily as a cat, and had no doubt watched me, puzzled by my stillness, till my first decided motion brought him out on the jump. But I am wandering away from the small boy getting his first lessons in the woods, and learning that the important thing is to hold perfectly still.

Later, when eight or nine years old, I went alone day after summer day to the wild berry-pastures. When my big pail would hold no more, I would make a bowl by bashing in the top of my hat, and fill it to the brim with luscious blueberries. These with a generous slice of bread made an excellent lunch, which I always ate within sight of a bird's nest, or the den of a fox, or some other abode of life that I had discovered in the woods. And again, as I sat quiet in the solitude, the birds and small animals might be led by curiosity to approach as fearlessly as when I was too small to harm them. Now a vixen, finding me too near her den and cubs, would squall at me impatiently, like a little yellow dog with a cat's voice; or again, a brooding bird that objected to my scrutiny would first turn her tail to me, and presently come round again, and finally get mad and flutter about my head, scolding loudly to chivvy me away. So it often happened that one had nearer or happier or more illuminating glimpses of wild life in that small hour of rest than would be possible in a month of roaming the woods with gun or collecting-box.

Once as I was eating my lunch under the pines, meanwhile watching a den I had found to see what might come out of it, a crow sailed in on noiseless wings and lit so near me that I hardly dared wink for fear he would notice the motion. My first thought was that he was nest-robbing (a crow is very discreet about that business), but he appeared rather to be listening, cocking his head this way or that; and from a lazy *hawing* in the distance I concluded he was satisfying himself that his flock was occupied elsewhere and that he was quite alone. Presently he hitched along the branch on which he stood and glided off to the crotch of a pine-tree, where he began to uncover what was hidden under a mat of brown needles. The first thing he took out was a piece of glass, which sparkled with rainbow colors in a stray glint of sunshine. Then came a bit of quartz with more sparkles, a shell, a silvery buckle, and some other glistening objects which I could not make out. He turned his treasures over and over, all the while croaking to himself in a pleased kind of way; then he put them all back, covered them again with needles, and slipped away without a sound. Having kept tame crows, I knew that they are forever stealing and hiding whatever bright objects they find about the house; and here in the pine woods was a thing to indicate that wild crows, perhaps all of them, have the same covetous habit.

Another day, a heavenly day when the budding woods were vocal and life stirred joyously in every thicket, I took a jews'-harp from my pocket and began to twang it idly. No, there was nothing premeditated in the act. I had been roving widely, following the winds or the bird-calls till a sunny opening invited me to rest, and had then fingered the music-maker with no more purpose than the poet's boy, who "whistled as he went for want of thought." The rhythmic, nasal twanging was a sound never heard in that place before or since, I think, and the first to come hurriedly to investigate was a bright-colored warbler, whose name I did not know; nor did I care to know it, feeling sure that by some note or sign he would presently suggest a name for himself, which would please me better than the barbarous jargon I might find in a bird-book. The alert little fellow lit on a branch within three feet of my face, turning his head so as to view me with one eye or the other when I kept quiet, or chirping his indignation when I twanged the jews'-harp. Next came a jay, officious as the town constable; then more birds, half concealing their curiosity under gentle manners; and a squirrel who had no manners at all, scolding everybody and scurrying about in a fashion which seemed dangerous to his excited head.

As I watched this little assembly, which seemed to be asking, "What's up? What's up?" the meaning of it suddenly dawned on me like a surprising discovery. When I entered the opening I knew simply that birds or beasts would draw near if they found me quiet; before I left it I had found the explanation: that all the wood folk are intensely curious, as curious as so many human gossips, but without any of their malice; that by inner compulsion they are drawn to any strange sight or sound, as a crowd collects when a man cuts a caper or throws a fit or raises a whoop or looks up into the air, or does anything else out of the ordinary. When you appear in the quiet woods every bird or beast within sight or hearing is agog to know about you; they are like the Nantucket-Islanders, who named their one public hack the "Who's Come?" Because you are a stranger, and what you do is none of their business, they are all the more interested in you and your doings; you come to them with all the charm of the unknown, the unexpected; and they will gratify their curiosity, fearlessly and most pleasantly, so long as you know how to stimulate or play upon it and to hold still while enjoying it.

All that is natural enough, as natural as life; but it is not written in any book of natural history, and it came to me that day as a wonderful discovery. It suggested at once the right way to study birds or beasts, as living creatures; it has since led to many a fascinating glimpse of the wood-folk comedy, and to a lifelong pleasure which is too elusive to be set down in words. At the bottom of it, I suppose, is the fact that in every wild or natural creature is something, at once mysterious and familiar, which appeals powerfully to your interest or sympathy, as if you saw a faint shadow of your other self, or caught a fleeting memory of that vanished time when you lived in a child's world of wonder and delight.

From the beginning, therefore, I met all birds and animals in a child's impersonal way; which, strangely enough, ascribes personality to every living thing, yes, and honors it. These inquisitive little rangers of the wood or the berry-pasture, shy and exquisitely alert, were all individuals like myself, each one seeking the joy of life in his own happy way. My only regret was that I was too clumsy, too obtrusive, too ignorant of the way of the wild, and so frightened many a timid bird or beast that I would gladly have known.

All this, too, is perfectly natural; the instinctive attitude of a child, as of an animal, is one of curiosity rather than of fear or destruction. If left to his natural instincts, a child meets every living creature with a mixture of shyness or timidity and bright interest; he becomes an enemy of the wild, learning to frighten and harry and kill, not from nature but from the evil example of his elders. I could prove that beyond a peradventure, I think, if this were the place; but there is no need of any man's demonstration. Go yourself to the big woods at twilight, leaving custom behind you; go alone and unarmed; hear that rustle of leaves, that tread of soft feet which brings you to an instant halt; see that strange beast which glides out into the trail and turns to look at you with luminous eyes. Then quickly examine your own mental state, and you will know the truth of a man's natural or instinctive attitude toward the mystery of life.

Unfortunately the wild birds and beasts near our home have learned that man is unnatural, a creature to be feared, and their curiosity has given place to another motive. The young still display their natural bent freely; but the old have heard too many of our guns, have been too often disturbed by our meddlesome dogs or worthless cats, have suffered too much at the hands of outrageous egg-collectors or skin-collectors to be any longer drawn to us when we go afield. As you go farther away from civilization it becomes easier to play on the animals' native curiosity; in the far North or the remote jungle, or wherever man is happily unknown, they still come fearlessly to investigate you, or to stand quiet, like the ptarmigan, watching with innocent eyes as you pass them by. In the intermediate regions, which are harried by sportsmen for a brief period in the autumn and then left to a long solitude, the animals are wild or tame according to season; and it has seemed to me, not always but on occasions, that in some subtle way they distinguish between man and man, taking alarm at the first sniff of a hunter, but stopping to show their interest in a harmless woods-rover.

This last is a mere theory, to be sure, and to some it may appear a fanciful one; but it rests, be assured, upon repeated experience. Thus, I came once at evening to a camp of hunters who were in a sorry plight. They were in a good deer country, and had counted largely on venison to supply their table; but for more than a week they had tasted no meat, and they were very hungry. The deer were wild as hawks, they assured me. They had hunted every day; but because of the game's wildness and the dry weather, which made the leaves rustle loudly underfoot, it had proved impossible to approach near enough for a shot— all of which made me think that, if you want to see game, you should leave your gun at home. I had met about a dozen deer that day; most of them were within easy range, and a few of them stood with questioning eyes while a man might have made ready his camera and taken a picture of them.

The very next morning, and within a mile of the hunters' camp, I witnessed a familiar but fascinating display of deer nature. At sunrise I approached a bog, bordering a stream where a few good trout might be found, and on the edge of the opening stood a doe and her well-grown fawn, not twenty yards away. The fawn, a little buck with the nubs of his first antlers showing, threw up his head as I appeared, and in the same instant I dropped to the ground behind a mossy log. No whistle or sound of alarm followed the action; so I scraped a mat of moss from the log, put it on for a bonnet, and cautiously raised my head.

The old doe was still feeding; the buck stood like a living statue, his whole attention fastened on the spot where I had disappeared. He had seen something, he knew not what, and was waiting for it to show itself again. When my bonnet appeared his eyes seemed to enlarge and flash as he caught the motion. Without changing his footing, for his surprise seemed to have rooted him in the ground, he began to sway his body to left or right, stretching his head high or dropping it low, taking a dozen graceful attitudes in order to view the queer bit of moss from different angles. Then he slowly raised a fore foot, put it down very gently, raised it again, stamped it down hard. Getting no response to his challenge, he sidled over to his mother, still keeping his eyes fastened on the log. At his touch or call she lifted her head, pointing her nose straight at me, as if he had somehow told her where to give heed.

It was a wonderful sight, the multicolored bog spread like a rug at the feet of the glorious October woods, and standing on the crimson fringe of it these two beautiful creatures, demanding with flashing eyes who the intruder might be. For a full minute, while I held motionless, the doe kept her eyes steadily on the log; then, "Nothing there, little buck; don't worry," she said in her own silent way, and went to feeding once more.

"But there is something; I saw it," insisted the little buck, nudging his mother by swinging his head against her side. That was the first and only time, in that quick swing, when he took his eyes from what attracted them. The doe looked a second time, saw nothing uncommon, and had turned to feed along the edge of the opening when the little buck recalled her in some way. "Can't you see it, that white thing like a face under the moss?" he was saying. "There! it moved again!"

The mother, whose back was turned to me, twisted her head around as if to humor him, and to interest her I swayed the moss bonnet to and fro like a pendulum. At that she whirled, surprise written large on her, and I dropped my head, leaving her staring. When I looked again both deer were coming nearer, the mother ahead, the fawn holding back as if to say, "Careful now! It's big, and it's hiding just behind that log." So they drew on warily, stopping to stamp a fore foot, and every time they challenged I gave the bonnet an answering wag. When they were so near that I knew they must soon distinguish my eyes from the moss, I sank out of sight. I was listening for their alarm-call or for the thud of their flying feet when a gray muzzle slid over the log, and I laid my hand fair on the mother's cheek before she bounded away.

Now there was nothing strange or new in all that; on the contrary, it was very much like what I had observed in other inquisitive deer. The only surprising part of the comedy was that the doe, though she had felt the touch of my hand and no doubt smelled the man behind it, stopped short after a few jumps and turned to stare at the log again. That she was still curious, still unsatisfied, was plain enough; what puzzles me to know is, whether she would have acted in the same way if one of the hungry hunters had been waiting in my shoes for the chance or moment to kill her.

IX

I T is easy, much easier than you think, to get close to wild birds and beasts; for after you have met them a few times in the friendly, impersonal way I have tried to describe, two interesting traits appear: the first, that they do not see you clearly so long as you hold still; the second, that even their keen noses lose track of you after you have been quiet for a little time.

The eye is a weak point in all animals I have chanced to observe, which apparently depend less on sight than on any other sense—so far as safety goes, that is. In gratifying their curiosity they seem to be all eyes. At other times they will catch an abrupt or unusual motion quickly enough; but they are strangely blind to any motionless object however large or small. Repeatedly when I have been sitting quiet, without concealment but with "neutral" clothes that harmonize with the soft woods colors, I have known deer, moose, caribou, bear, wolf, fox, lynx, otter, alert beasts of every kind, to approach within a few yards, giving no heed till a chirp or a slight motion called their attention. Then they would whirl upon me in astonishment, telling me by their attitude that till then they had not noticed or suspected me. Almost invariably at such times animals of the deer family would come a step nearer, their heads high, their eyes asking questions; but beasts of prey after one keen look would commonly drop their heads, as if I were of no consequence, and slyly circle me to get my wind. In either event success or a better view of the animal depended on just one condition, which was to hold absolutely still. So long as I met that condition, none of these wary beasts seemed to have any clear notion what they were looking at.

Occasionally, indeed, their lack of discernment almost passes belief. One winter day, while crossing a frozen lake in Ontario, I noticed a distant speck moving on the snow, and stopped in my tracks to watch it. The speck turned my way, drew near and changed into an otter, who came rollicking along in his merry way, taking one or two quick jumps on his abbreviated legs and a long slide on his ample belly. As the air was dry and very still I had no fear of his nose, which is not as sensitive as many others (perhaps because of the peculiar valve or flap which closes it tight when an otter swims under water); but his eyes and other senses are extraordinarily good, and it seemed impossible that he should overlook a man standing erect on the snowy ice, as conspicuous as a fly in the milk. So I watched the approach with lively interest, wondering how Keeonekh would act in comparison with other members of his weasel family when he found himself near me, whether he would dart away like a fisher, or ignore me like a mink, or show his teeth at me like a little stoat.

On he came, confidently, as an otter travels, giving no heed to the enemy in his path, till he halted with a paw resting on one of my snowshoes and began to wiggle his broad muzzle, as if he found something in the air which he did not like. For several moments he hesitated, sniffing here, listening there, looking sharply about the lake, into the near-by woods, everywhere except up into my face, and then went on as he had been heading, leaving a straight trail behind him. No man can tell what was in his head, and a very intelligent head it is; but his action seemed to say that he did not see me when he passed literally under my nose. To him I was merely a stump, one of a dozen that projected here or there above the ice near the shore.

Such an incident would be merely freakish if it happened once; but it happens again and again, becoming almost common or typical, when a man stands motionless in the presence of other birds or beasts. Twice in the big woods has the Canada lynx, a dull beast in comparison with the otter, passed me with an unseeing stare in his wild eyes. And I have crouched in the snow on a treeless barren while a band of caribou filed past, so near that I could see the muscles ripple under their sleek skins and hear the *click-click* of their hoofs as they walked. The greater part of the herd did not even notice me; the rest threw a passing glance in my direction, one halting as if he had a moment's doubt, and went on without a sign of recognition.

The eyes of birds are keener, as a rule; but it is still a question with me how much or how little they see of what is plain as a frog on a log to human vision. An owl has excellent eyes, which are at their best in the soft twilight; yet once as I sat quiet in the dusk a horned-owl swooped and struck at a motion of my head, not seeing the rest of me, one must think, since he is quick to take alarm when a man appears. Another owl showed even less discernment in that he overlooked me, head and all, at a yard's distance and tried to get his game out from under my feet.

A little dog had followed me that day, keeping out of sight till we were far from home, when he showed himself in a waggish way, as if he knew I would not have the heart to tie him up in that lonely place, as he deserved. All day long he had a vociferous and a "bully" time making a nuisance of himself, stirring up a hornets' nest in every peaceful spot, chasing deer out of sight with a blithe rowdydow, swimming out to the raft on which I was fly-fishing, jumping in to get tangled in my landing-net when I reached for a big trout—in twenty ways showing that his business was to take care of me, though he was no dog of mine.

Late in the afternoon, as I rested beside the homeward trail, the little dog rambled off by himself, still looking for trouble, like all his breed. Presently there was a yelp, a scurry, a glimpse of broad wings swooping, and back came the trouble-seeker like a streak, his eyes saying, "Look at this thing I brought you!" and his ears flapping like a pair of wings to help him along. Over him hovered a big barred-owl, grim as fate, striking, missing, mounting, swooping again, brushing me with his wings as he whirled around my head. Between my heels and the log on which I was sitting my protector wedged himself securely; the owl with a vicious snapping of his beak sailed up into an evergreen and made himself invisible. In a moment or two the little dog came out and was wagging his tail mightily over the adventure when the owl slanted down on noiseless wings and struck a double set of claws into him. Then I interfered, rising to my feet; and then, for the first time I think, the owl saw me as something other than a stump and vanished quickly in the spruce woods.

Hawks likewise have marvelous eyes for all things that move; but I began to question the quality of their vision one day when I was watching a deer, and a red-shouldered hawk lit so near me that I reached out a hand and caught him.

Another afternoon I came upon a goshawk, keenest of all the falcons, that had just killed a grouse in the tote-road I was following. He darted away as I came round a bend; but thinking he might soon return, for he is a bold and a persistent kind of pirate, I entered the woods at a swift walk, as if going away. Then I worked cautiously back to the road through a thicket, and waited on a log in deep shadow, some fifty yards above where the grouse lay undisturbed. Luckily I had a rifle, an accurate little twenty-two, often carried as medicine for "vermin," and I intended to kill the goshawk at the first chance. He is the viking among birds, and as such has a romantic interest; but wherever he appears he is a veritable pest, the most destructive of the hungry hordes that come down from the North to play havoc with our game.

For a long time the goshawk hovered about, sweeping on tireless wings high above the trees; but though I was quiet enough to deceive any bird or beast, he held warily aloof. Once he disappeared, remaining so long away that I was beginning to think he had wearied of the game of patience, when I heard an eery call and saw him wheeling over the road again. His absence became clear a little later when he perched on a blasted pine, far out of range, where he remained watching for fifteen or twenty minutes, his only motion being an occasional turning of the head. That he was hungry and bound to have his own was plain enough; the puzzle was why he did not come and get it, for it seemed highly improbable that he would notice a motionless figure at that distance. I think now that he was sailing high over my head when I re-entered the trail; that he knew where I was all the time, not because he saw me on the log, but because he did not see me go away. Before sunset he headed off toward the mountain, going early to roost with all his kind.

Dawn found me back on the old road, as much to test the matter of vision as to put an end to the game-destroyer. As a precaution I had changed clothes, and now shifted position, avoiding the log because the wary bird would surely take a look at it before coming down. My stand was a weathered stump beside the road, against which I sat on a carpet of moss, without concealment of any kind. At a short distance lay the grouse, a poor crumpled thing, just as he had wilted under the hawk's swoop.

Thus a half-hour or more passed comfortably. A gorgeous cock-partridge sauntered into the open, saw me when I nodded to him, and went slowly off with that graceful, balancing motion which a grouse affects when he is well satisfied with himself. Then, as the sun rose, there was a swift-moving shadow, a rustle of pinions; the goshawk swept down the road in front of me and lit beside his game. He was a handsome bandit, in full adult plumage; his gray breast was penciled in shadowy lines, while his back was a foggy blue, as if in his northern home he had caught the sheen of the heavens above and of rippling waters beneath his flight. His folded wings stood out squarely from his shoulders with an impression of power, like an eagle's. There was something noble in his poise, in his challenging eye, in the forward thrust of his fierce head; but the spell was broken at the first step. He moved awkwardly, unwillingly it seemed; his great curving talons interfered with his footing when he touched the earth.

This time, instead of a rifle, there was a trim shotgun across my knees. The hawk was mine whether he stood quiet or leaped into swift flight, and feeling sure of him now I watched awhile, wondering whether he would break up his game with his claws, as some owls do, or tear it to pieces with his hooked beak. For a moment he did neither, but stood splendidly alert over his kill. Once he turned his head completely around over either shoulder, sweeping his piercing glance over me, but seeing nothing unusual. Then he seized his game in one foot and struck his beak into the breast, making the feathers fly as he laid the delicate flesh open. When I found myself weakening, growing sentimental at the thought that it was his last meal, his last taste of freedom and the wild, I remembered the grouse and got quietly on my feet. Though busy with his feast, he caught the first shadow of a motion; I can still see the gleam in his wild eyes as he sprang aloft.

I thought him beyond all harm as he lay on his back, one outstretched wing among the feathers of his victim; but he struck like a flash when I reached down for him carelessly. "Take that! and that! and remember me!" he said, driving his weapons up with astonishing force, a force that kills or paralyzes his game at the first grip. Four of his needle-pointed talons went to the bone, and the others were well buried in the flesh of my arm. The old viking had been some time with his ancestors before I pried him loose.

As for the sense of smell, on which most animals depend for accurate information, I have tried numerous experiments with deer, moose, bear and other creatures to learn how far they can wind a man, and how their powers compare one with another. There is no definite answer to the problem, so baffling are the conditions of observing these shy beasts; but you are in for some surprises, at least, when you attempt to solve it in the open. You will learn, for example, that when a gale is blowing the animals are more at sea than in a dead calm; or that in a gusty wind you can approach them about as easily from one side as from another. Such a wind rolls and eddies violently, rebounding from every hill or point or shore in such erratic fashion that the animals have no means of locating a danger when they catch a fleeting sniff of it. It is for this reason, undoubtedly, that all game is uncommonly wild on a windy day: the constant motion of leaves or tossing boughs breeds confusion in their eyes, and the woodsy smells are so broken by cross-currents that they cannot be traced to their source. So it has happened more than once on a gusty day that a deer, catching my scent on the rebound, has whirled and rushed straight at me, producing the momentary illusion that he was charging.

With a steady but not strong wind blowing in their direction, I have seen deer become alarmed while I was yet a quarter-mile away; this on a lake, where there was nothing to interfere with the breeze or the scent. On the burnt lands or the open barrens I have seen bear and caribou throw up their heads and break away while I was even farther removed. In a light breeze the distance is much shorter, varying from fifty to two hundred yards, according to the amount of moisture in the air. On days that are still or very dry, or when the air is filled with smoke from a forest fire (the latter soon inflames all sensitive nostrils), the animals are at sea again, and depend less on their noses than on their eyes or ears.

Another surprising thing is, that the animal's ability to detect you through his sense of smell is largely governed by your own activity or bodily condition. Thus, when a man is perspiring freely or moving quickly, his scent is stronger and travels much wider than when he is sauntering about. But if a man sits absolutely quiet, a clean man especially, no animal can detect him beyond a few feet, I think, for the reason that a resting man is like a resting bird or beast in that he gives off very little body scent, which remains on the ground close about him instead of floating off on the air currents. Even when the trees are tossing in a gale there is little stir on the ground, not in the woods at least, and the closer you hold to Mother Earth the less likelihood is there of any beast smelling you.

All ground-nesting birds depend for their lives on this curious provision of nature. Were it not for the fact that practically no scent escapes while they are brooding their eggs, very few of them would live to bring forth a family in a wood nightly traversed by such keen-nosed enemies as the fox and the weasel. My old setter would wind a running grouse or quail at an incredible distance, and would follow him by picking his scent from the air; but I have taken that same dog on a leash near the same birds when they were brooding their eggs, and he could not or would not detect them unless he were brought within a few feet, or (a rare occurrence) unless a creeping ground-breeze blew directly from the nest into his face.

The same provision guards animals, such as deer and caribou, which build no dens but leave their helpless young on the ground. Two or three times, after finding a fawn in the woods, I have tested his concealment by means of my young dog's nose; and I may add that Rab will point a deer as stanchly as he points a grouse or woodcock, for he is still in the happy, irresponsible stage when everything that lives in the woods is game to him. So long as the fawn remains motionless where his mother hid him, the dog must be almost on top of him before pointing or showing any sign of game. But if the little fellow runs or even rises to his feet at our approach (fawns are apt to do this as they grow older), the dog seems to catch the scent *after* the first motion; he begins to cat-foot, his nose up as in following an air trail, and steadies to a point while he is still many yards away from where the fawn was hiding.

The nose of a wolf is keener than that of any dog I ever knew; yet I once trailed a pack of wolves that passed within sixteen measured feet of where two deer were sleeping in a hole in the snow. The wolves were hunting, too, for they killed and partially ate a buck a little farther on; but the trail said that they had passed close to these sleeping deer without detecting them.

As for the man-scent, you may judge of that by the violent start or the headlong rush when an animal catches the first alarming whiff of it. If he passes quietly on his way, therefore, you may be reasonably sure he has not smelled you. To the latter conclusion I have been forced many times when I have been watching in the woods, sitting quiet for hours at a stretch, and a deer or bear or fox, or some other beast with nose as keen as a brier, has passed at a dozen yards' distance without a sign to indicate that he was aware of me. Some of these animals came much nearer; so near, in fact, that I was scary of a closer approach until I had called their attention to what lay ahead of them.

So long as you are seen or suspected, you need have little fear of any wild beast (only the tame or half-tame are dangerous), but a brute that stumbles upon you in an unexpected place or moment is always a problem. Nine times out of ten he will fall all over himself in his haste to get away; but the tenth time he may fall upon you and give you a mauling. Moose, for example, are apt to strike a terrible blow with their fore feet, or to upset a canoe when the jack-light approaches them; not to attack, I think, at least not consciously, but in blind panic or to ward off a fancied enemy. So when I have watched from the shore of a lake and a moose came swinging along without noticing me, I have risen to my feet or thrown my hat at the big brute when he was as near as I cared to have him. And more than once, after a tremendous start of surprise, he has come nearer with his hackles up as soon as he got over the first effect of my demonstration. Yet when I am roaming the woods that same brute will catch my scent at from two to five hundred yards, and rush away before I can get even a glimpse of him.

That the same surprising sense-limitation is upon deer and other game animals may be inferred from the following experience, which is typical of many others. I was perched among some cedar roots on the shore of a pond, one September day, watching a buck with the largest antlers I have ever seen on one of his kind. I had been some time quiet when he glided out to feed in a little bay, on my right; and my heart was with him in the wish that he might keep his noble crown through the hunting season, for his own pleasure and the adornment of the woods and the confusion of all head-hunters. There was no breeze; but a moistened finger told of a faint drift of air from the lake to the woods.

As I watched the buck, there came to my ears a crunching of gravel from the opposite direction, and two deer appeared on the point at my left, heading briskly down into the bay. They passed between my outstretched feet and the water's edge, where the strip of shore was perhaps three yards wide; then they turned in my direction, seeing or smelling nothing, went slowly up the bank and halted at the edge of the woods to the right and a little behind me, so close that I dared not move even my eyes to follow them. I measured the distance afterward, and found that from their hoof-marks to the cedar root against which I rested was less than eight feet. Imperceptibly I turned for another look, and saw both deer at attention, their heads luckily pointed away from me. They were regarding the big buck intently, as if to question him. They showed no alarm as yet; but they were plainly uneasy, searching the forest on all sides and at times turning to look over my head upon the breathless lake. Every nervous action said that they found something wrong in the air, some hint or taint or warning which they could not define. So they moved alertly into the woods, halting, listening, testing the air, using all their senses to locate a danger which they had passed and left behind them.

From such experiences one might reasonably conclude that, like the brooding grouse or the hidden fawn, a motionless man gives off so little scent that the keenest nose is at fault until it comes almost within touching distance. If any further proof is needed, you may find it when you sleep in the open, and shy creatures draw near without any fear of you. By daylight deer, bear and moose are extremely timid; they rarely come within eyeshot of your camp, and they vanish at the first sniff which tells them that you have invaded their feeding-grounds. But when you are well asleep the same animals will pass boldly through your camp-yard; or they will awaken you, as they have many times awakened me, when you are tenting or sleeping under the stars by some outlying pond. If you lie quiet, content to listen, the invading animal will move freely here or there without concern; but no sooner do you begin to stir, however quietly, than he catches the warning scent, and a thudding of earth or a smashing of brush tells the rest of the story.

I recall one night, cloudy and very still, when I slept under my canoe on a strip of sand beside a wilderness lake. The movement of an animal near at hand awoke me. In the black darkness I could see nothing; but somehow I knew he was big, and aside from the crepitation of the sand, which I plainly heard, I seemed to feel the brute near me. For a moment there was a pause, a dead silence; then came a thump, a rattlety-bang; the canoe shook as something hit the lower end of it, and the creature moved away. There was nothing to be done without eyes, so I snuggled the blanket closer and went to sleep again. In the morning there were the tracks of a moose, a bull as I judged from the shape of his feet, to say that he had come down the shore at a fast walk, halted, stepped over the stern of the canoe, and went on without hastening his pace.

That was odd enough; but more surprising were some tracks on the other side, between the bow of the canoe and the woods. Very faint and dainty tracks they were, as if a soft pad had touched the sand here and there in an uneven line; but they told of a fox who had come trotting along under the bank, and who had passed in the night without awakening me. That neither he nor the moose had smelled the sleeping man, or nothing alarming in him at least, is about as near to certainty as you will come in interpreting animal action.

There is another and not wholly unreasonable hypothesis which may help to explain such phenomena; namely, that it is not the scent of man but of excitement, anger, blood-lust or some other abnormal quality which alarms a wild animal. It sounds queer, I know, to say that anger can be smelled; but it is more than probable that anger or fierce excitement of any kind distils in the body a kind of poison which is physical and sensible. Such excitement certainly weakens a man, clogging his system with the ashes of its hot fires; and there is no reason why it should not smell to earth as well as to high heaven.

You have but to open your eyes and expand your nostrils for some evidence of this matter. Bees when angered give off a pungent odor, which is so different from the ordinary smell of the hive that even your dull nose may detect the change of temper. The same is true of even cold-blooded reptiles. When you find a rattler or a black-snake squirming in the sun, you can smell him faintly at a few yards' distance. Now stir him up with a pole, or pin him to the earth by pressing a forked stick with short prongs over his neck. As the snake becomes enraged he pours off a rank odor, very different from the musky smell that first attracted your notice, and it travels much wider, and clings to your clothes for an hour afterward. It is not only possible but very likely, therefore, that strong emotions affect the bodies of all creatures in a way perceptible to senses other than sight. If so, one man who is peaceable and another who is angry or highly excited may give off such different odors that a brute with sensitive nostrils may be merely curious about the one and properly afraid of the other.

That wild animals instinctively fear the scent of humanity, as such, is probably not true. The notion arises, I think, from judging the natural animal by those we have made unnatural by abuse or persecution. Whenever man penetrates a wild region for the first time he finds, as a rule, that the animals have little fear of him, the tameness of wild game having been noted with surprise by almost every explorer. It has been noted also, but without surprise, by saints and ascetics who "for the greater glory of God" have adopted a life of solitude and meditation, and who have often found the birds or beasts about their hermitage to be quite fearless of them, and receptive of their kindness. Not till the abundant flocks and herds of a new region have been harried and decimated by senseless slaughter do the survivors begin to be fearful and unapproachable, as we unfortunately know them. Yet even now, no sooner do we drop our persecution and assume a rational or humane attitude than the wild ducks come to the boat landing of a winter hotel, deer feed at our haystacks, and bears come in broad daylight to comfort themselves at our garbage-cans. Such things could hardly be if the fear of man were an age-old or instinctive inheritance.

Nearer home, on any farm bordering the wilderness, you may see wild deer feeding quite tamely about the edges of the cleared fields all summer. I recall one such farm in Maine, where the owner had fifteen acres of green oats waving over virgin soil—a glorious crop for me, but for him an occasion of lamentation. You could go through that field at any hour before six in the morning or after six at night and find a dozen deer with a moose or two making themselves at home. The owner's cattle were kept out by a rail fence; but the moose simply leaned against the fence and went through, while the nimble deer sailed over the obstruction like grasshoppers. On all such farms the deer have the scent of man almost constantly in their nostrils, and they are simply watchful, running when you approach too near, but turning after a short flight to have a look at you. At times you may see them feeding when the scent of laborers or fishermen blows fairly over them. But when October comes, and the law is "off," and wild-eyed hunters appear with guns in their hands and death in their thoughts, then the same deer quickly become as other and wilder creatures, rushing off in alarm at the first sniff of an enemy. The fact and the changed action are evident enough; the only interesting question is, To what extent does the smell of man change when he changes his peaceable ways?

Two or three times I have had opportunity to test the effect of the human scent in another way, the first time being when I had the good luck to see a natural child and a natural animal together. The child, a baby girl just beginning to toddle, was making a journey by means of a comfortable Indian *paukee* on my back, and I had left her in an opening beside a portage trail while I went back to my canoe for a thing I had forgotten. While I was gone, three deer sauntered into the opening. They saw the baby, and were instantly as curious about her as so many gossips, a little spotted fawn especially. The baby saw them, and began creeping eagerly forward, calling or "crowing" as she went. The deer saw and heard and smelled her every moment; yet they walked around her with springy steps, now on this side, now on that, showing a world of curiosity in their bright eyes, but never a sign of fear.

From a distance I watched the lovely scene, kindling at the beauty of it, or feeling a bit anxious when I saw the sharp feet of the old doe a little too near the sunny head or the outstretched hands. Then an eddy of wind from the mountain got behind me and whirled over the deer. They caught the scent and were away with a wild alarm-call, their white flags flying, and the baby waving by-by as they vanished in the woods.

Quite naturally, therefore, when a sensitive animal runs away from me, I find myself thinking that perhaps it is not the smell of humanity but of some evil trait or quality which frightens him. I first laid down this hypothesis after meeting a strange, childlike man, who had a passion for roaming by himself in the fields or woods. White men, after a puzzling acquaintance, would tap their heads or call him crazy; an Indian would look once in his eyes and say, very softly, "The Great Spirit has touched him." He was all gentleness, without a thought or possibility of harm in his nature. He was also without fear, and perhaps for this reason he inspired no fear in others. When he appeared in the woods, singing to himself, the animals would watch him for a moment, and then go their ways quietly, as if they understood him. What would happen if a race of such men lived near the wood folk must be left to the imagination.

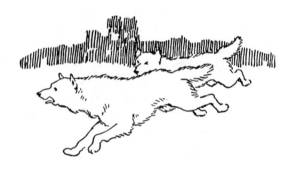

My Pond: a Symphony of the Woods

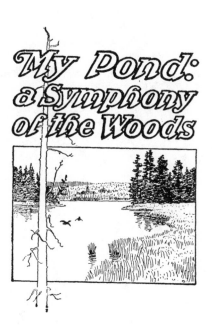

The Trail

X

To reach my pond you must leave your canoe on the shore of Sungeegamook, the home lake, and go eastward through the big woods. Yonder is the landing, that bank of green topped by "everlasting" and blue asters, with a cleft like an arched doorway in the forest behind it. A rugged jack-pine leans out over a bit of shingle, as if to indicate a good place to beach your canoe, and there is something curiously alive, almost sentient, in its attitude. The old tree seems to watch your approach; through its leaves runs a low murmur of welcome as you step ashore.

Entering the woods (and because you are alone, and therefore natural, something in their dim aisles, their mysterious depths, their breathing silence, makes you go gently) you find yourself in an old logging-road, once a garish symbol of man's destructiveness, but growing yearly more subdued, more beautiful, since Nature began her work of healing. The earth beneath your feet, the restful earth which the lumbermen left torn by iron tools or rent by dynamite, has again put on her soft-colored garments. Feathery beds of fern push boldly into the road from shadowy places; wild grasses fill all its sunny openings with their bloom and fragrance; and winding down through shade or sunshine comes a trail made by the feet of deer and moose. Already these timid animals have adopted the forgotten road as a runway; you may meet them here when you return in the evening twilight.

Everywhere beside the trail are old marks of the destroyer. Noble maples or cedars that were centuries growing have been slashed down, dismembered, thrust aside to decay, and all because they stood in the way of a lumber-boss who thought only of getting his cut of spruce down to the lake. To look upon such trees, dead and shorn of their beauty, is to feel pity or indignation; but Nature does not share your feeling, being too abundant of life and resource to waste any moment in regret. Already she is upbuilding what man has torn down. Glaring ax-wounds have all disappeared under bandages of living moss; every fallen log has hidden its loss under a mantle of lichen, soft and gray, which speaks not of death but of life renewed.

Where the sun touches these prostrate giants a blush of delicate color spreads over them. See, it deepens as you look upon it curiously, and you examine it to find a multitude of "fairy-cups" on slender stems, each lifting its scarlet chalice to the light. Very soft and inviting seats they offer, yielding to your weight, sending up an odor as of crushed herbs; but do not accept the invitation. If you must halt to rest or to enjoy the stillness, sit not down on one of these mossy logs, but before it at a little distance, and let its blended colors be to your eye what the wind in the pine is to your ear, or the smell of hemlock to your nostrils. Then will all your senses delight in harmony, their natural birthright, while you rest by the way.

Where the old road winds about the end of a ridge, avoiding every steep pitch, young balsams are crowding thickly into it; where it turns downward to the lowlands, quick-growing alders claim it as their own; and as you leave the lake far behind it begins to divide interminably, each branch breaking into smaller branches, like the twigs of a tree as you trace them outward. The twig ends with a bud in clear space; but the farther or landward end of a logging-road dwindles to a deer-path, the path to a rabbit-run, and the run vanishes in some gloomy cedar swamp or trackless thicket where is no outlook on any side.

It is in such places, while you puzzle over another man's road instead of keeping your own trail straight, that you are most apt to get lost. Coming back you need have no fear of going astray, since all these trails lead to the main road, and thence downhill to the lake; but going forward it is well to steer clear of all branch roads, which lead nowhere and confuse the sense of direction.

Leaving the road behind, therefore, and heading still eastward, you cross a ridge where the hardwoods stand, as their ancestors stood, untouched by the tools of men. Immense trunks of beech or sugar-maple or yellow birch tower upward wide apart, the moss of centuries upon them; far overhead is a delicate tracery of leaves, a dance of light against the blue, and over all is the blessed silence.

Beyond the ridge the ground slopes downward to a uniform level. Soon the moss grows deeper underfoot, with a coolness that speaks of perpetual moisture. The forest becomes dense, almost bewildering; here a "black growth" of spruce or fir, there a tangle of moosewood, yonder a swale where impenetrable alder-thickets make it impossible to hold a straight course. Because all this growth is useless to the lumberman, there is no cutting to be seen; but because I have passed this way before, instinctively following the same course like an animal, a faint winding trail begins to appear, with a bent twig or a blazed tree at every turn to give direction.

As you move forward more confidently, learning the woodsman's way of looking far ahead to pick up the guiding signs before you come to them, the dim forest suddenly brightens; a wave of light runs in, saying as it passes overhead that you are near an opening. As if to confirm the message, the trail runs into a well-worn deer-path, which looks as if the animals that used it knew well where they were going. Clumps of delicate young larches spring up ahead; between them open filmy vistas, like windows draped in lace, and across one vista stretches a ribbon of silver. A few more steps and—there! my little pond is smiling at you, reflecting the blue deeps of heaven or the white of passing clouds from its setting of pale-green larch-trees and crimson mosses.

And now, if you are responsive, you shall have a new impression of this old world, the wonderful impression which a wilderness lake gives at the moment of discovery, but never again afterward. As you emerge from cover of the woods, the pond seems to awaken like a sleeper. See, it returns your gaze, and on its quiet face is a look of surprise that you are here. Enjoy that first awakening look; for there is more of wisdom and pleasure in it, believe me, than in hurrying forth blindly intent on making a map or catching a trout, or doing something else that calls for sight to the neglect of insight. All sciences, including chartography and angling, can easily be learned by any man; but understanding is a gift of God, and it comes only to those who keep their hearts open.

Your own nature is here your best guide, and it shows you a surprising thing: that your old habitual impressions of the world have suddenly become novel and strange, as if this smiling landscape were but just created, and you were the first to look with seeing eyes upon the glory of it. It tells you, further, if you listen to its voice, that creation is all like this, under necessity to be beautiful, and that the beauty is still as delightful as when the evening and the morning were the first day. This dance of water, this rain of light, this shimmer of air, this upspringing of trees, this blue heaven bending over all—no artist ever painted such things; no poet ever sang or could sing them. Like a mother's infinite tenderness, they await your appreciation, your silence, your love; but they hide from your description in words or pigments.

Finally, in the lowest of whispers, your nature tells you that the most impressive and still most natural thing in this quiet scene is the conscious life that broods silently over it. As the little pond seems to awaken, to be alive and sentient, so also does that noble tree yonder when you view it for the first time, or that delicate orchid wafting its fragrance over the lonely bog. Each reflects something greater than itself, and it is that greater "something" which appeals to you when you enter the solitude. Your impressions here are those of the first man, a man who found many beautiful things in a garden, and God walking among them in the cool of the day. Call the brooding life God or the Infinite or the Unknown or the Great Spirit or the Great Mystery—what you will; the simple fact is that you have an impression of a living Being, who first speaks to you in terms of personality that you understand.

So much, and more, of eternal understanding you may have if you but tarry a moment under these larches with an open mind. Then, when you have honored your first impression, which will abide with you always, you may trace out the physical features of my pond at leisure. Just here it is not very wide; your eye easily overlooks it to rest with pleasure on a great mound of moss, colored as no garden of flowers was ever colored, swelling above the bog on the farther shore. On either hand the water sparkles wider away, disappearing around a bend with an invitation to come and see. To the left it ends in velvety shadow under a bank of evergreen; to the right it seems to merge into the level shore, where shadow melts with substance in a belt of blended colors. A few yards back from the shore groups of young larches lift their misty-green foliage above the caribou moss; they seem not to be rooted deep in the earth, but to be all standing on tiptoe, as if to look over the brim of my pond and see their own reflections. Everywhere between these larch groups are shadowy corridors; and in one of them your eye is caught by a spot of bright orange. The spot moves, disappears, flashes out again from the misty green, and a deer steps forth to complete the wilderness picture with the grace of life.

Such is my pond, hidden away in the heart of a caribou bog, which is itself well hidden in dense forest. Before I found it the wild ducks had made it a summer home from time immemorial; and now, since I disturb it no more, it is possessed in peace by a family of beavers; yet I still think of it as mine, not by grace of any artificial law or deed, but by the more ancient right of possession and enjoyment. A hundred lakes by which I have tented are greater or more splendid; but the first charm of any wilderness scene is its solitude, and on these greater lakes the impression of solitude may be broken by the flash of a paddle-blade in the sun, or the *chuck* of an ax under the twilight, or the gleam of a camp-fire through the darkness. But here on my pond you may know how Adam felt when he looked abroad: no raft has ever ruffled its surface; no ax-stroke or moan of smitten tree has ever disturbed its quiet; no camp-fire has ever gleamed on its waters. Its solitude is still that of the first day; and it has no name, save for the Indian word that came unbidden at the moment of finding it, like another Sleeping Beauty, in the woods.

Do you ask how I came to find my pond? Not by searching, but rather by the odd chance of being myself lost. I had gone astray one afternoon, and was pushing through some black growth when an alarm rose near at hand. A deer whistled loudly, crying *"Heu! heu! heu!"* as he jumped away, and on the heels of his cry came a quacking of flushed ducks.

Till that moment I thought I knew where I was; but the quacking brought doubt, and then bewilderment. If a duck tells you anything in the woods, he tells you of water, plenty of it; but the map showed no body of water nearer than Big Pine Pond, which I had fished that day, and which should be three or four miles behind me. Turning in the direction of the alarm, I soon broke out of the cover upon a caribou bog, a mysterious expanse never before suspected in that region, and before me was the gleam of water in the sunshine. "A pond, a new one, and what a beauty!" I thought with elation, as I caught its awakening look and feasted my eyes on its glory of color. Then I gave it an Indian name and hurried away; for I was surely off my course, and the hour was late for lingering in strange woods. Somewhere to the west of me was the home lake; so westward I headed, making a return-compass of bent twigs, till I set my feet in a

branch of the old logging-road. And that chance trail is the one I have ever since followed.

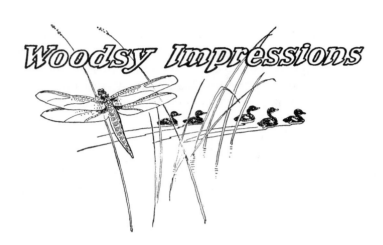

XI

NEXT morning I returned to explore my find at leisure. One part of that exploration was to go completely around the bog, to learn its guiding landmarks and compass-bearings; but an earlier and better part was to sit quietly beside my pond to hear whatever it might have to say to me. If that last sounds fanciful, remember that many things are voiceless in this world, but few are wholly dumb. Of the numberless ponds that brighten the northern wilderness, some were made by beavers, others by flood or glacier or earthquake, and no two of them tell the same story or make the same impression. They are like so many unspoiled Indians, whom we regard from a distance as being mysteriously alike, but who have different traditions, ideals, personalities, and even different languages.

I know not what the spell of any lonely place may be when you make yourself part of it; I only know that it stirs one strangely, like the flute note of a wood-thrush or a song without words. Though I never met with an adventure on my little pond, never cast a fly to learn whether any trout lurked in its waters, never thought of firing a shot at its abundant game, yet season after season I returned to it expectantly, and went away satisfied. Such a pond has a charm of its own, a spell which our forebears sought to express in terms of nymphs or *puckwudgies* or water-sprites. It grows a better crop than trout, attracts a finer game than deer or water-fowl, and you can seldom visit it without learning something new about your natural self or the wood folk or the friendly universe.

Thus, it happens on a day when you are waiting beside your pond, or wending your way to it, that a moose or a fox or a dainty grouse appears unexpectedly near you; and instantly, without thought or motive, you "freeze" in your tracks or, if you are not seen, shrink deeper into the shadow for concealment. The action is natural, involuntary, instinctive, precisely like the action of a young deer under similar circumstances; but when it is over you understand it, and smile at finding yourself becoming more and more like other natural creatures,—going softly, that is, making yourself inconspicuous without trying or knowing how, and having no thought of harm to any bird or beast, but only of watching him or gauging his course while remaining yourself unseen. Only by some such method can you learn anything worth knowing about a wild animal: books describe, naturalists classify and sportsmen kill him; but to understand him you must be a sharer of his quiet ways.

Comes another day, a day when you are in love with solitude itself, when you learn with surprise that a man is never lonely when alone in the woods; that ideals may be quite as companionable as folks; and that around you in a goodly company are beauty, peace, spacious freedom and harmonious thoughts, with a hint also, to some minds, of angels and ministers of grace. The Attendant Spirit of "Comus," the Ariel of "The Tempest," the good fairies of all folk,—these are never understood in the town, nor in the woods unless you enter them alone.

At a later time, and with a thrill of great wonder, you may discover the meaning of silence, and of the ancient myth of a lovely goddess of silence; not the dead silence of a dungeon, which may roar in a man's ears till it deafens him or drives him mad, but the exquisite living silence of nature, a silence which at any moment may break into an elfin ringing of bells, or into a faintly echoing sound of melody, as if stars or unseen beings were singing far away.

This impression of melody is often real, not illusory, and may be explained by the impact of air-currents on resonant shells of wood, hundreds of which fall to humming with the voice of 'cellos and wind-harps; but there is another experience of the solitude, more subtle but none the less real, for which only the psychologist will venture to give an accounting. Once in a season, perhaps, comes an hour when, no matter what your plans or desires may be, your mind seems intent on some unrelated affair of its own. As you hurry over the trail, you may be thinking of catching a trout or stalking a buck or building a camp or getting to windward of a corporation; meanwhile your subconscious mind, disdaining your will or your worry, is busily making pictures of whatever attractive thing it sees,—radiant little pictures, sunshiny or wind-swept, which shall be reproduced for your pleasure long after the important matters which then occupied you are clean forgotten.

Here is the story of one such picture, a reflection, no doubt, of the primitive trait or quality called place-memory, which enables certain animals or savages to recognize any spot on which their eyes have once rested.

One late afternoon, years after I had found my pond, I crossed the mountain from distant Ragged Lake, heading for the home lake by a new route. There was no trail; but near the foot of the western slope of the hills I picked up an old lumber road which seemed to lead in the right direction. For a time all went well, and confidently; but when the road dipped into an immense hollow, and there showed signs of petering out, I followed it with increasing doubt, not knowing where I might come out of the woods or be forced to spend the night. As I circled through a swale, having left the road to avoid a press of alders that filled it, an ash-tree lifted its glossy head above a thicket with a cheery "Well met again, pilgrim! Whither away now?"

It was a surprising hail in that wild place, suggestive of dreams or sleep-walking; but under the illusion was a grain of reality which brought me to an instant halt. After passing under thou-

sands of silent trees all day, suddenly here was one speaking to
me. And not only that, but wearing a familiar look, like a face
which smiles its recognition of you while you try in vain to place
it. Where, when had I seen that tree before? No, impossible! I
had never before entered this part of the vast forest. Yet I must
have seen it somewhere, or it could not now stir a familiar
memory. Nonsense! just a trick of the imagination. I must hurry
on. Thus my thoughts ran, like a circling hare; and all the while
the ash-tree seemed to be smiling at my perplexity.

The man who ignores such a hint has much to learn about
woodcraft, which is largely a subconscious art; so I sat down to
smoke a councilpipe with myself and the ash-tree over the matter. No sooner was the mind left to its own unhampered way
than it began to piece bits of a puzzle-picture deftly together;
and when the picture was complete I knew exactly where I was,
and where I might quickly find a familiar trail. Eight years
before, in an idle hour when nothing stirred on my pond, I had
explored a mile or so beyond the bog to the south, only to find
a swampy, desolate country without a trail or conspicuous landmark of any kind. It was while I passed through this waste, seeking nothing in particular and returning to my pond, that the
mind took its snapshot of a certain tree, and preserved the picture so carefully, so minutely, that years later the original was
instantly recognized. Many similar ash-trees grew on that flat,
each with its glossy crown and its gray shaft flecked by dark-
green moss; what there was in this one to attract me, what outward grace or inward tree-sprite, I have not yet found out.

Another subconscious record seems to have been made for
beauty alone, with its consequent pleasure, rather than for util-
ity. As I watched my pond one summer morning, intent on
learning what attracted so many deer to its shores, the mind
apparently chose its own moment for making a perfect picture,
a masterpiece, which should hang in its woodsy frame on my
mental wall forever. The sky was wondrously clear, the water
dancing, the air laden with the fragrance of peat and sweet-
scented grass. Deer were slow in coming that morning, and
meanwhile nothing of consequence stirred on my pond; but
there was still abundant satisfaction in the brilliant dragon-flies

that balanced on bending reeds, or in the brood of wild ducks that came bobbing out like young mischief-makers from a hidden bogan, or even in the face of the pond itself, as it brightened under a gleam of sunshine or frowned at a passing cloud or broke into a laugh at the touch of a cat's paw wind. Suddenly all these pleasant minor matters were brushed aside when a bush quivered and held still on the farther shore.

All morning the bushes had been quivering, showing the silvery side of their leaves to every breeze; but now their motion spoke of life, and spoke truly, for out from under the smitten bilberries came a bear to stand alert in the open. The fore part of his body was lifted up as he planted his paws on a tussock; his massive head was thrust forward as he tried to penetrate the far distance with his near-sighted eyes. He was not suspicious, not a bit; his nose held steady as a pointing dog's, instead of rocking up and down, as it does when a bear tries to steal a message from the air. A moment he poised there, a statue of ebony against the crimson moss; then he leaped a bogan with surprising agility, and came at his easy, shuffling gait around a bend of the shore. Opposite me he sat down to cock his nose at the sky, twisting his head as he followed the motion of something above him, which I could not see,—a hornet, perhaps, or a troublesome fly that persisted in buzzing about his ears. Twice he struck quickly with a paw, apparently missing the lively thing overhead; for he jumped up, rushed ahead violently and spun around on the pivot of his toes. Then he settled soberly to his flat-footed shuffle once more, and disappeared in a clump of larches, which seemed to open a door for him as he drew near.

For me that little comedy was never repeated, though I saw many another on dark days or bright; and the last time I visited my pond I beheld it sadly altered, its beauty vanished, its shores flooded, its green trees stark and dead. Unknown to me, however, the mind had made its photographic record, and always I see my pond, as on that perfect day, in its setting of misty-green larches and crimson bog. Again its quiet face changes, like a human face at pleasant thoughts, and over it comes to me the odor of sweet-scented grass. The sunshine brightens it; the clouds shadow it; brilliant dragon-flies play among its bending reeds; the same brood of ducklings glides in or out from bogan to grassy bogan; and forever the bear, big and glossy black, goes shuffling along the farther shore.

Larch-trees and Deer

XII

ONE of the subtler charms of my pond, a thing felt rather than seen, was a certain air of secrecy which seldom left it. In every wilderness lake lurks a mystery of some kind, which you cannot hope to penetrate,—a sense of measureless years, of primal far-off things, of uncouth creatures dead and gone that haunted its banks before the infancy of man; but on this little pond, with its sunny waters and open shore, the mystery was always pleasant, and at times provoking, as if it might be the place where an end of the rainbow rested.

Though small enough to give one a sense of possession (one can never feel that he owns a big lake, or anything else which gives an impression of grandeur or sublimity), my pond had a mischievous way of hinting, when you were most comfortable, that it was hiding a secret; that it might show you, if it would, a much better scene than that you looked upon. It was shaped somewhat like an immense pair of spectacles, having two lobes that were flashing bright, with a narrow band of darker water between; and, what with its bending shores or intervening larches, you could never see the whole of it from any one place.

So, like eyes that hide their subtlest lights of whim or fancy under glasses, it often seemed to be holding something in reserve, something which it would not reveal unless you searched for it. After watching awhile from one beautiful or restful spot, you began to feel or imagine that some comedy was passing unseen on the other half of the pond; and though you resisted the feeling at first, sooner or later you crept through the screen of larches to know if it were true.

On every side of the pond save one, where a bank of evergreen made velvet shadows intermingled with spots of heavenly blue, the shores were thickly spread with mosses, which began to color gloriously in midsummer, the colors deepening as the season waned, till the reflecting water appeared as the glimmering center of a gorgeous Oriental rug. Along the edges of this rug, as a ragged fringe, stood groups of larches in irregular order,—little fairylike larches that bore their crown of leaves not as other trees bear them, heavily, but as a floating mist or nebula of sage green. Like New England ladies of a past age they seemed, each wearing a precious lace shawl which gave an air of daintiness to their sterling worth. When the time came for the leaves to fall, instead of rustling down to earth with a sound of winter, mournfully, they would scamper away on a merry wind, mingling their fragrance with that of the ripened grass; and then the twigs appeared plainly for the first time, with a little knot or twist in every twig, like toil-worn fingers that the lace had concealed.

Here or there amid this delicate new growth towered the ruin of a mighty tamarack, or ship-knee larch, such as men sought in the old clipper-ship days when they needed timbers lighter than oak, and even tougher to resist the pressure of the gale or the waves' buffeting. Once, before the shipmen penetrated thus far into the wilderness, the tamaracks stood here in noble array, their heads under clouds, beckoning hungry caribou to feed from the lichens that streamed from their broad arms above the drifted snow; now most of them are under the moss, which covered them tenderly when they fell. The few remaining ones stand as watch-towers for the hawks and eagles; their broken branches make strange sepia drawings of dragon-knots and hooked beaks on the blue sky. A tiny moth killed all these great larches; the caribou moved northward, leaving the country, and the deer moved in to take possession.

This and many other stories of the past my little pond told me, as I watched from its shores or followed the game-trails that were spread like a net about its edges. Back in the woods these trails wandered about in devious fashion, seeking good browse or easy traveling; while here or there a faint outgoing branch offered to lead you, if your eyes were keen, to the distant ridge where a big buck had his daily loafing-place. On the bog the trails went more circumspectly, uniting at certain places in a single deep path, a veritable path of ages, which was the only path that might safely be followed by any creature with more weight than a fox. The moment you ventured away from it the ground began to shiver, to quake alarmingly, to sink down beneath your feet. Only a thin mat of roots kept you afloat; the roots might anywhere part and drop you into black bottomless ooze, and close forever over your head. A queer place, one might think, for heavy beasts to gather, and so it was; but the old caribou-trails or new deer-paths offered every one of them safe footing.

At first these game-trails puzzled me completely, being so many and so pointless. That they were in constant use was evident from the footprints in them, which were renewed almost every morning; yet I never once saw a deer approach the water to drink or feed. Something else attracted them; a highway from one feeding-ground to another, it might be, or the wider outlook which brings deer and caribou out of their dim woods to sightly places; but there was no certainty in the matter until the animals themselves revealed the secret. One day, when a young buck passed my hiding-place as if he were going somewhere, I followed him to the upper or southern end of the pond. There he joined four other deer, which were very busy about a certain spot, half hidden by low bushes, a couple of hundred yards back from the shore. And there they stayed, apparently eating or drinking, for a full half-hour or more.

When the deer were gone away, I went over and found a huge spring, to which converged a dozen deep trails. Like the hub of an immense wheel it seemed: the radiating paths were the spokes, and somewhere beyond the horizon was the unseen rim. From the depths of the spring came a surprising volume of clear, coffee-colored water, bubbling over joyously as it leaped from the dark earth into the light, and then stealing quietly away under bending grasses to keep my pond brim full. Around the spring the earth was pitted by the feet of deer, and everywhere about its edges were holes lapped in the peat by eager tongues. Here, beyond a doubt, was what called so many animals to my pond,—a mineral spring or salt-lick, such as we read about in stories of pioneer days, when game was everywhere abundant, but such as one now rarely finds.

After that happy discovery I shifted my blind to another larch with low-drooping branches, beneath which one might rest comfortably and look out through a screen of lace upon a gathering of the deer. They are creatures of habit as well as of freedom; and one of their habits is to rest at regular intervals, the hours being hard to forecast, since they vary not only with the season of lengthening or shortening days, but also each month with the changes of the moon. Thus, when the moon fulls and weather is clear, deer are abroad most of the night. At dawn they seek their day-beds, instinctively removing far from where they have left their scent in feeding; and during the day they are apt to remain hidden save for one brief hour, when they take a comforting bite here or there, giving the impression that they eat now from habit rather than from hunger. As the moon wanes they change their hours to take advantage of its shining; and on the "dark of the moon" they browse only in the early part of the night, then rest many hours, and have two periods of feeding or roaming the next day.

Such seems to be the rule in the North, with plenty of exceptions to keep one guessing,—as in the November mating-season, when bucks are afoot at all hours; or during a severe storm, which keeps deer and all other wild animals close in their coverts.

Because of this regularity of habit at irregular hours, the only certainty about the salt-lick was that the animals would come if one waited long enough. As I watched expectantly from my larch bower, the morning shadows might creep up to me, halt, and lengthen away on the other side, while not a deer showed himself in the open. Then there would be a stir in the distant larches, a flash of bright color; a doe would emerge from one of the game-trails, hastening her springy steps as she neared the spring. As my eyes followed her, noting with pleasure her grace-ful poses, her unwearied alertness, her frequent turning of the head to one distant spot in the woods where she had left her fawn, there would come another flash of color from another trail, then two or three in a flecking of light and shadow, till half a dozen or more deer were gathered at the lick, some lapping the mud eagerly, others sipping, sipping, as if they could never have enough of the water. After a time they would slip away as they had come, singly or in groups; the spring would be desert-ed, and one could never tell how many hours or days might pass before another company began to gather.

However eager for salt they might be, the deer came or went in that mysteriously silent way of theirs, appearing without warning in one trail, or vanishing down another without a sound to mark their passing. Now and then, however, especially if one watched at the exquisite twilight hour, a very different entrance might be staged on the lonely bog,—a gay, prancing "here I come: get out of the way" kind of entrance, which made one glad he had stayed to witness it. On the slope of the nearest ridge your eye would catch an abrupt motion, the upward surge of a bough or the spring-back of a smitten bush; presently to your ears would come a rapid thudding of earth, or a *sqush, sqush, sqush* of water; the larches would burst open and a buck leap forth, flourishing broad antlers or kicking up mad heels as he went gamboling down the game-trail. If other deer were at the spring, they would throw up their heads, set their ears at the dancing buck, take a last quick sip from the spring, and move aside as he jumped in to muzzle the mud as if famished. For it was the mud rather than the water which first claimed his atten-tion, no doubt because it held more of the magic salt. He often gave the impression, as he approached in high feather, that he had been tasting the stuff in anticipation and could hardly wait to get his tongue into it.

The first time I saw that frisky performance I went over to taste the mud for myself, but found little to distinguish it from the mud of any other peat-bog. The water from the spring was wholesome, with a faint taste of something I could not name; and I drank it repeatedly without learning its secret. That it held a charm of some kind, which chemistry might reveal, was evident from the fact that deer came from miles around to enjoy its flavor. Some of the trails could be traced clear across the bog to distant ridges and a broken country beyond; and in following these trails, to learn what creatures used them and where they came from, I repeatedly came upon a deer asleep in his day-bed. Whether the animals couched here before drinking at the spring, or after drinking, or "just by happentry" I could not tell.

Once the sleeper was a buck with noble antlers. He was resting beside a great log on the edge of an opening, half surrounded by dense fir thickets. I speak of him as asleep; but that is mere habit of speech or poverty of language. Of a score of wild birds or beasts that I have found "asleep" in the woods, not one seemed to lose touch with the waking world even for an instant. The buck's eyelids were blinking, his head nodding heavily; yet all the while his feet were curled in readiness for an instant jump; and somehow those expressive feet gave the impression of being as wide awake as a squirrel. Occasionally as I watched him, fascinated by the rare sight, his head would drop almost to the ground, only to be jerked up with an air of immense surprise; then the sleepy fellow would stare in a filmy, unseeing, "who said I was asleep" kind of way at a little tree that stood in the opening. The stare would end with a slow closing of the eyelids, and in a moment he would be nodding again.

Black Mallards

XIII

NEXT to the deer, the wild ducks were the chief attraction of my pond. Indeed, they might well be placed first, since they were always at home there, and much of the time engaged in one or another of the little comedies that make ducks the most amusing of all birds. Eight summers in succession, and again after an interval of two years, I found my pond occupied by a pair of black mallards with their brood; and I fancied, since migratory birds return to the place of their birth, and their nestlings after them, that one of the pair was the lineal descendant of ducks that had held the place in undisputed possession for tens of thousands of years. Here was a succession, modest like all true nobility, which made the proud family trees of *Mayflower* folk or English kings or Norman barons look like young berry-bushes in the shade of a towering pine.

Until late midsummer the family had the pond all to themselves. Never a stranger-duck appeared to share or challenge their heritage; while day after day the mother watched over the little brood as they fed or played or learned the wild-duck signals. Like our dogs, every manner of beast or bird has its own tribal ways or customs, some of which do not appear in the young until they begin to roam abroad or to mingle with their kind. So, as I watched the brood emerge from down to pin-feathers, there would come a red-letter day when two of them, meeting as they rounded a grassy point, would raise their wings as if in salutation; and a later day when, the pin-feathers having grown to fair plumage, their young cheepings or whistlings would change to a decided *quack*.

Thereafter their talk was endlessly entertaining, if one took the trouble to creep near enough to appreciate its modulations, expressive of every emotion between drowsiness and tense alarm; for it cannot be heard, except as a meaningless sound, beyond a few yards. The little hen-ducks got on famously, having the mother's quacking as a model; but male ducks cannot or will not learn to quack, and since a male voice was rarely heard on the pond at this season, each little drake was a law unto himself, and made a brave show of his liberty. Climbing on a tussock, as if for more room, he would stretch his wings, make odd motions with his neck, and finally pump out a funny *wheekle, wheekle,* as if he had swallowed a whistle.

Meanwhile the old drake and father of the family was seldom about; only two or three times did I see him enter the pond, stay a brief while, and then wing away over the tree-tops in the direction of a larger lake, some three miles to the eastward. On that lake there was never a brood of young ducks, so far as I could learn; but when trout-fishing there I often surprised the drake, at times taking precious care of his own skin in solitude, again clubbing sociably with three or four other drakes, who had run away each from a family and the cares thereof on some other lonely pond.

As the summer waned, a new sound of quacking, joyous and exultant, would greet me when I drew near my pond. Creeping to my blind under the larches, I would find a second brood making merry acquaintance with the family I had watched over; then a third and a fourth company of strangers, as young ducks of all that region began to traffic about in preparation for the autumn flight. A little later the flocks fairly reveled in sociability, gathering here or there with increasing numbers, till on a late-

September day I might find my pond deserted, the owners being on a visit elsewhere, or I might catch breath at sight of so many ducks that I could not accurately count them or distinguish one brood from another.

At such a time my little pond seemed to awaken, to shed its silence like a garment, to put on its most animated expression, as at a happy festival or family reunion. The air was never still from the gabble of meeting groups (probably all more or less related), or from the resounding *quank, quank, quank* of some old gossip who went about proclaiming her opinion to the whole company. Everywhere the still water was broken into undulating wakes as the drakes swept grandly over it, with that rhythmic, forward-and-back motion of their heads which is like duck poetry,—a motion that is not seen when the birds are feeding, but only when they are well satisfied with themselves or their audience. Through the shadows under the bank glided knots or ribbons of young birds which had not yet quite satisfied their appetites, some exploring every crevice for ripened seeds, others tip-tilting their tails to the blue sky as they probed the bottom for water-bugs and other tit-bits. In an open space a solitary hen-duck bobbed and teetered ecstatically, dipping the fore part of her body under, then heaving it up quickly so as to send the cleansing water in a foamy wave over her back and wings. Here or there on a tussock stood a quiet group of the splendid birds, oiling their glossy feathers, setting a wing-cover just right, or adding some other last touch to an elaborate toilet before settling down for a nap.

The glassy water reflected every form, color, motion of these untroubled ducks as in a glass, doubling the graceful effect. Around them stretched the gloriously colored bog; and beyond the bog were the nebulous-green larches, the somber black growth and the lifting hills, on which autumn had laid its golden touch. Truly a beautiful sight, a sight to make the heart of hunter or naturalist tremble with expectancy as he fingered his gun. I have known that trembling, that expectancy; but there was greater pleasure, perhaps greater freedom also, in leaving the happy comedy undisturbed.

Because of its solitude, its utter wildness, my pond seemed to be the chosen resting-place of the flocks on an autumn day (they feed or travel mostly by night), and perhaps for the same reason the ducks that frequented it were among the wildest creatures I have ever tried to stalk. A black mallard is not an easy bird to outwit at any time or place; but here some magic mirror or sounding-board seemed to supplement his natural eyes or ears. The slightest unnatural voice or appearance, the snap of a twig or the quiver of a leaf or the glimpse of a face in the larches, would send a flock away on the instant; and sometimes, when I was sure no sound or motion of mine had broken the perfect quiet, they would take wing in such incomprehensible fashion as to leave me wondering what extra sense had warned them of danger.

Several times in the course of a summer, when I wanted to observe the little duck family more nearly, or to learn the meaning of some queer play that I could not understand from a distance, I would creep out of the larches unseen, worming my way along a sunken deer-path, and stopping whenever heads were turned in my direction. One might think it an easy matter to approach any game by such methods; yet almost invariably, before I could be safe behind a bush or a tuft of grass at the water's edge, the old mother-duck would become uneasy, like a deer that catches a vague hint of you floating far down the wind. That she could not see or hear me was certain; that she could not smell me I had repeatedly proved; nevertheless, after searching the shores narrowly she would stretch her neck straight up from the water, as if attentive to some wireless message in the air.

A wild duck does not take that alert attitude unless she is suspicious; and a curious thing was, that though the mother was silent, uttering never a word, the young would crouch and remain motionless wherever they happened to be. Suddenly, as if certain of danger but unable to locate it, the mother would spring aloft to go sweeping in wide circles over the bog. She seemed to know it by heart, every pool and bump and shadow of it; and when her keen eyes picked up an unfamiliar shadow on a certain deer-path she would come at it with a rush, whirling over it in an upward-climbing spiral till she became sure of me, as of something out of place, when she would speed away with a warning note over the tree-tops. If the young were strong of wing, they would follow her swiftly, giving wide berth to the deer-path as if she had told them beware of it; but if they did not yet trust themselves in the air, they would skulk away, their heads down close to the water, and hide in one of the grassy bogans of the pond, where because of the quaking shore it was impossible to come near them.

Once, when the mother left in this way, I waited till the duck-lings had been some minutes hidden before creeping back to my blind in the larches. An hour or more passed in the timeless quiet; while the water became as glass under the afternoon sun, and a deer moved near the hidden brood without flushing them or even bringing a head up where I could see it. Then the mother returned, calling as she came; and the first thing she did was to circle warily over the same deer-path, stretching her neck down for a close inspection. "Aha! that thing is gone, but where?" she said in every line and motion of her inquisitive head or pulsating wings, as she sped away to find the answer.

Twice she circled the bog, her eyes searching every cranny and shadow of it. From her high flight she slanted straight down and pitched fair in the middle of the pond, where for some moments she sat motionless, her head up, looking, listening,—a perfect image of alertness in the midst of wildness. Satisfied at last that no trouble was near, she turned to the shore with a low call; and out of the bogan pell-mell rushed the little ones, splashing, cheeping, half lifting themselves with their tiny wings as they scurried over the water to join the mother. For a full hour I had kept my glasses almost continuously on that bogan; then with divided attention I cast expectant glances at it when I heard the mother's incoming note, the whish of her wings as she circled the bog and the splash as she took the water; but not till the right signal came did I see a motion or a sign of life from the hidden brood.

The pond was shaped, as we have noticed, like a pair of spec-tacles; and a favorite place for the autumn flock to rest or preen or sleep was at the bend between the two lobes. Down into that bend ran a screen of alder-bushes, the only good cover between woods and water on the entire pond; and it was so dense that a cat could hardly have crept through it without making a disturbance. That was one reason, I suppose, why the ducks felt safe at the outer end of the tangle: they could see everything in front or on either side, and hear anything that moved behind them.

One day, when the shore at this bend was freshly starred by ducks' feet and littered with feathers, showing that a large flock had just left the roost, I began at the fringe of larches and cut a passageway, a regular beaver's tunnel, down the whole length of the alder run, making an end in a point of grass, where the water came close on three sides. One had to consider only the birds' keen ears, the alder screen being so thick that not even a duck's eye could penetrate it; therefore I smoothed the way most carefully, leaving no stick below to crack under my weight, and no branch reaching down to rustle or quiver as I crawled beneath it. When the tunnel was well finished I left the pond to its solitude a few days, thinking that the birds would surely notice some telltale sign of my work, some fresh-cut stick or wilted bough that my eyes had overlooked, and be wary of the alders for a little time.

And why such pains to get near a bird, you ask, since one might better observe or shoot him from a comfortable distance? Oh, just a notion of mine, an odd notion, which can hardly be appreciated till one has proved it in the open. As you can seldom "feel" the quality of a stranger while he remains even a few yards away, so with any wild bird or beast: there is an impression arising from nearness, from contact, which cannot be had in any other way; and that swift impression, which is both physical and mental, a judgment as it were of the entire nature, is often more illuminating than hours of ordinary observation or speculation.

Such an impression is not new or strange, or even modernly psychological. On the contrary, it is the simplest matter in the world of sense, I think, and perhaps also the surest. Most animals have a significant way of touching their noses to one of their own kind at meeting; not to smell him, as we imagine (they can smell him, or even his tracks, at a distance), but in order to receive a more intimate or convincing message than the sense of smell can furnish. Likewise, a man naturally pats the head of a dog, or fingers an object after minutely scanning it with his eyes; and in this instinctive action is the ancient touch of recognition. Touch is the oldest and most universal of the bodily senses, sight, smell, taste and hearing being later specializations thereof; by it the living creature first became aware of a world outside of self; and to it we all return for verification of our sense impressions. Therefore it happens most naturally that, despite warning signs or penalties, thoughtless men will put their hands into the bear or monkey cage, where animals are no longer natural or to be trusted, and our children must be forever lectured, or sometimes spanked, for handling things which they have been told to let alone.

Besides, when one is very near a strange bird or beast, one becomes vaguely conscious of an extra sense at work,—that real but uncatalogued sense-of-presence (to coin a name for it) which makes two persons in a room aware of each other at every instant, even while both are absorbed in quiet work or reading. The "feel" of the same room when one occupies it alone is very different; and the difference may help to explain why gregarious animals are uncomfortable, uneasy, unless they are near their own kind,—near enough, that is, not simply to hear or see them but to feel their bodily presence. A herd-animal is always restless, and often sickens, if his herd is not close about him. The same mysterious sense (mysterious to us, because we do not yet know the organ through which it works) often warns the solitary man in the woods or in the darkness that some living creature is near him, at a moment when his eyes or ears are powerless to verify his impression.

But that is another and more subtle matter, familiar enough to a few sensitive persons and natural woodsmen, but impossible of demonstration to others; you cannot explain color to a man born blind. The simple answer is, that for my own satisfaction I wanted to touch one of the wary birds of my pond, as I had before touched eagle and crow, bear and deer, and many another wild creature in his native woods. Such was the notion. In other places I had several times tried to indulge it; but save in one instance, when I found a winter flock weakened by hunger,

I had never laid my hand fairly on a black mallard when he had the free use of his wits and wings.

When I returned to my pond, and from a distance swept my glasses over it, the water was alive with ducks; never before had I seen so many there at one time. Single large birds, the drakes undoubtedly, were moving leisurely over the open spaces. Groups of five or six, each a brood from some neighboring pond, were gliding in an exploring kind of way under the banks or through the weed-beds; and scattered along the shore at the end of the alder run were wisps or companies of the birds, all preening or dozing with an air of complete security. Here at last was my chance, my perfect chance, I told myself, as I carefully marked one brood standing at the tip of the grassy point where my tunnel ended.

More carefully than ever I stalked a bear, I circled through the black growth, crept under the fringe of larches, and entered the alder run unobserved. Inch by inch I wormed along the secret passageway, flat to the ground, not once raising my head, hardly daring to pull a full breath, till, just as I emerged from the alder shade into the grass, a gamy scent in my nose and a low gabble in my ears told me that I was almost near enough, that the birds were all around me, and that for the rest of the way I must move as a shadow.

From under my hat-brim I located the gabblers, a large family of black mallards outside the fringe of grass on my left. They were abreast of me, not more than five or six feet away. I had not marked these birds when I began my stalk; they were hidden in a tiny cove or bend of the shore, and had it not been for their voices I would surely have crept past without seeing them. At the mouth of the cove was a single tussock, on which stood the mother-duck, wavering between dreams and watchfulness as the sunshine poured full upon her, making her very sleepy. On the bare earth beneath her the others were getting ready for a nap, so near that I could see every motion, the settling of a head, the blink of an eyelid. Occasionally through the tangle of grass stems came the penetrating gleam of their eyes,—marvelously bright eyes, alert and intelligent.

For several minutes I held motionless, still flat to the ground, listening to the sleepy talk, admiring the mottled-brown plumage of a breast or the bar of brilliant color drawn athwart a sooty wing. All the while my nose was trying to get in a warning word, telling me to give heed that the ducky odor which flowed in waves over the whole point was different from this strong reek, as of a disturbed nest, in the near-by grass; but my eyes were so occupied that I paid no attention to other senses. As the duck on the tussock at last settled down to sleep and I worked my toes into the earth for a noiseless push forward, there was a slight but startling motion almost at my shoulder. A neck was raised and twisted sleepily, as if to get the kink out of it; and the thrill of success ran over me as I made out another and nearer group of ducks. They were under the bank and the bending grass, where I had completely overlooked them. Every one was within reach; and every one I could see had his head drawn in or tucked away under his wing.

Slowly my left hand stole toward them, creeping forward in the deliberate fashion of a measuring-worm, first the fingers stretched, then the knuckles raised, then out with the fingers again. It would have been very easy to stroke or to catch one of the birds by a swift motion; but that was not what I wanted, and would have instantly spoiled the whole comedy. For the right effect, the hand must rest upon a duck before he was aware of it, so quietly that at first he would give his attention to the hand itself, not to the thing it came from. Then he would probably give it a questioning peck, examine it curiously, and finally grow indifferent to it, as other birds had done when I touched them from hiding. But here my head was too close to the ground, and my body too cramped for easy action. As my hand reached the edge of the bank, just over an unconscious duck, it ran into a tuft of saw-grass, which cut my fingers and rustled dangerously. To clear this obstruction I drew back slightly, lifted up a grain; and in my other ear, which was turned away, a resonant voice cried *Quock!* with a challenge that broke the tension like a pistol-shot.

Involuntarily I turned my head, just when I should have held most still; and so I lost my chance. There, at arm's-length on the other side of the point, a wild-eyed duck was looking over the bank, her neck stretched like a taut string, her olive-colored bill pointing straight at me. She never said another word, and had no need to repeat her challenge. All over the point and along the shore necks were stretched up from the grass; a dozen alert forms rose like sentinels from as many tussocks, and forty pairs of keen eyes were every one searching the spot at which the old hen-duck pointed her accusation.

For a small moment that tableau lasted, without a sound, without a motion; while one was conscious only of the tense necks, the pointing bills, the gleaming little eyes, each with its diamond-point of light; and then the old duck took wing. She did not even crouch to jump, so far as I could follow her motion; she simply went into the air like a rocket, shooting aloft as if hurled from a spring. As she rose, there was an answering rush of wings, *whoosh!* in my very ears, a surge as of smitten water in the distance; and in the same fraction of an instant every duck to the farthest ends of the pond was up and away in a wild tumult of quacking.

Only one of these birds had seen me, and that one probably had no notion of what she had glimpsed in the grass. It was a round thing with eyes, and it moved a second time,—that was enough for the old hen-duck, and the others did not stop to ask any questions.

Memories

XIV

Two full years passed before I returned to my pond on a sunny September day, in my mind's eye seeing it smile a welcome, hearing it cry, "Lo here! Lo there!" and planning, as I came down the silent trail, how I would accept all its invitations. First, the salt-lick must be spied out from a distance; and the examination would tell me whether to keep on down my own trail or, if the lick were occupied, to branch off by a certain game-path, which would lead me to the blind where I had so often watched the deer unseen. Next, I would have a restful look at a mound of moss swelling above the bog near a certain tamarack, which always showed the first blush of crimson in midsummer, and which became in autumn like a gorgeous bed of Dutch tulips, only more wondrously colored. Then I would look into the doorway under the larches, where my bear had disappeared. I always picked that out from a hundred similar doorways to watch or question it a moment, as if at any time the green curtain might open to let the bear out. For a curious thing about all woodsmen is this: if they see a buck or a bear or even a fox enter a certain place, they must forever afterward stop to have another expectant look at it.

From the bear's doorway my thoughts turned naturally to a little bogan of my pond, which was different from all the other bogans, because once a family of minks darted out of it and came dodging along the shore in my direction. Luckily I was close to the water at that moment. While the minks were out of sight under some bushes, I swung my feet over the bank and sat down in their path to wait for them.

In advance came the mother, looking rusty in her sunburnt summer coat, and she was evidently in a great hurry about something. The little ones, trailing out behind, were hard put to it to keep up the pace. She was fairly under me before she noticed a new scent in the air, which made her halt to look about for the meaning of it. Her neck was lifted, weasel-fashion, to thrice its ordinary length; at the end of it her pointed head swung like a vane to the bank, to the pond, to the bank again; while her busy nose wiggled out its sharp questions. Probably she had no notion of man, never having met the creature; neither did she associate the motionless figure above her with life or danger. She passed directly over one of my shoes, halted with her paws raised against the other, and scampered on as if she had no use for such trifles.

Before the little ones arrived I half turned to meet them, spreading my feet so as to leave a narrow passageway between the heels; and over this, as a cover, rested my hand, making a shadowy runway such as minks like. When the kits entered it, sleek and glossy and half grown, I touched them lightly on the neck, feeling the soft brush of fur and the ripple of elastic muscles as one after another glided under my finger, with no more concern than if it had been one of the roots among which they were accustomed to creep. But when the last one came I blocked the runway by placing a hand squarely across it, stopping him short in great astonishment. He sniffed at the obstruction, and his nose was like a point of ice as it wandered over my palm. Then he tried a finger with his teeth, wriggled under it to follow his leader, and the whole family disappeared in a twisting, snakelike procession around the next bend. These were *wild* animals, remember; and ounce for ounce there is no more "savage" beast in the woods than Cheokhes the mink.

As with birds or beasts, so also with the trees about my pond: somehow they seemed different from all other trees, perhaps because of more intimate association; for though all the cedars or hemlocks of a forest look alike to a stranger, no sooner do you spend days alone among them than you begin to have a curious feeling of individuality, of comradeship, of understanding even, as if they were not wholly dumb or insensate. It was inevitable, therefore, as I came down the trail, recalling this or that tree under which I had often passed or rested, that certain of them stood forth in memory as having given me pleasure or greeting in the lonely woods, just as certain faces emerge from the sea of faces in a crowd or a great audience of strangers, and instantly make one feel his kinship to humanity.

Foremost among these memorable trees was a great white-pine, to me the noblest of all forest growths, which stood on a knoll to westward of my pond, on the way to camp, and which always seemed to cry hail or farewell as I came or went. It had a stem to make one wonder, almost to make one reverent. Massive, soft-colored, finely reticulated it was; wide as the span of a man's arms, and rising near a hundred feet without knot or branch,—a glorious upspringing shaft, immensely strong, yet delicate in its poise as a lance in rest. From the top of the shaft rugged arms were stretched out above the tallest trees, and on these rested lightly as a cloud its crown of green. Like others that overtop their fellows, the old pine had paid the penalty of greatness. Whirlwinds that left lower trees untouched had stripped it of half its branches; lightning had leaped upon it from the clouds, leaving a spiral scar from crown to foot; but the wound which threatened its death was meanwhile its life, because the lumbermen, seeing the lightning's mark, had passed on and left the pine in its solitary grandeur.

When I first saw that tree I changed the trail so as to pass beneath it; and thereafter it was like a living presence, benign and friendly, beside the way. To lay a hand on its mighty stem, as one passed eastward in the early morning, was to receive an impression of renewed power,—a power which the scornful might attribute to imagination, the chemist to electrons or radio-activity, and the simple man to his Mother Nature. At evening, as one followed the dim trail homeward in the fading light, one had only to look up for a guiding sign; and there, solemn and still against the twilight splendor, was the crown of the old pine to give direction. Its very silence at such an hour was like the Angelus ringing. To halt beneath it, as one often did unconsciously, was to feel the spell of its age, its serenity, its peace; while harmonious thoughts came or went attuned to the low melody of the winds, crooning their vesper song far up among its green leaves. And, morning or midday or evening, to look up at the pine's lofty crown, which had tossed in the free winds that bore Pilgrim and Puritan westward with their immortal dream of freedom, was to be bound with stronger ties of loyalty to the fathers of my native state,— men of vision and imagination as well as of stern courage, who heard the pine booming out its psalm to the gale and instantly adopted it as their new symbol, stamping it on their coins or emblazoning it on their banners as an emblem of liberty. Never another symbol, whether dragon or eagle or lion, had so much majesty, or was so worthy of free men. The remembrance of it in any national crisis or call to duty sets the American heart beating to the rhythm of Whittier's "Pine-Tree":

Lift again the stately emblem on the Bay State's rusted shield,
Give to northern winds the Pine-Tree on our banner's tattered field.
Sons of men who sat in council with their Bibles round the board,
Answering England's royal missive with a firm "Thus saith the Lord,"
Rise again for home and freedom! set the battle in array!
What the fathers did of old time we their sons must do to-day.

Very different from the majestic pine was a little larch-tree, under which I often sat while watching the deer. As I came down the trail, after a year's absence, it would seem to lift its head and step forth from all the other larches, calling out cheerily: "Welcome once more! And why so long away? See, here is your old place waiting." And drawing aside the delicate branches, I would find the seat of dry moss and springy boughs, the back-rest, the open window with its drapery of lace,—everything just as I had left it.

Near this sociable young larch stood its dead ancestor, grim and silent, which the moths had killed; and this, too, seemed different from all other trees living or dead. On sunny days it threw a straight shaft of shadow over my blind; and the shadow moved along the ground from west to east, telling the creeping hours like a sundial. At the tip of the lofty stub a short branch thrust itself out at a right angle, and this served as the finger of my strange timepiece. When it rested on a bed of brimming pitcher-plants it pointed to the lunch in my pocket; when it touched the root of a water-maple it spoke of the home trail; and between, at irregular intervals, were a nanny-bush, a tuft of wild cotton and a shy orchid to mark the less important hours. Once, when I glanced at the slow-moving shadow, it was topped by a striking symbolic figure, and looking up quickly I found an eagle perched on the outstretched finger of my dial. After that the old tamarack had a new dignity in my eyes; it stood on an eagle's line of flight, one of his regular ways in crossing from mountain to lake, and from it the kingly-looking bird was wont to survey this part of his silent domain, the sun gleaming on his snow-white crest.

A stone's-throw behind my larch blind was a portly young fir, which I could never pass without a smile as it nodded to remind me that it was not like other firs. Thousands of these trees, crowding the northern forest, seem to be all grown on the same model, like peas in a pod; but this one had a character and a history to set it forever apart from its kind. And this is the tale which always passed silently between us when we met:

One day, as I watched some deer at the salt-lick, they suddenly became uneasy, looking and harking about as if for danger, and then vanished down the several game-trails. Not till they were gone did I notice that the air was ominously still, or understand the cause of the alarm: a tempest was coming, and the sensitive animals were away to cover before my dull senses had picked up the first warning sign. Soon the landscape darkened; the face of my pond became as I had never seen it before; thunder growled in the distance; coppery clouds with light flaming through them came rolling over the tree-tops; and all nature said, as plainly as a fire-bell, "Get to cover, and quickly!"

As I went back into the woods, seeking shelter, a few big drops hit the leaves like flails; then came a pause, still as death, and then the deluge. Ahead in the gloom I spied a young fir (never pick a tall tree, or a solitary tree, in a tempest of lightning) which thrust out a mass of feathery branches from a thicket of its fellows. "This for mine," I said as I dived under it, accompanied by a blinding flare of light and an ear-splitting crack—and almost ran against the heels of a buck that jumped out on the other side. By an odd chance, one in ten thousand, he had picked the same fir for shelter, and was no doubt thinking he had picked well when I came blundering in with the thunderbolt and drove him out into the downpour. "Hold on, old sport! Come back; it's your tree," I called after him, feeling as if I had stolen a child's umbrella; but he paid no attention.

Thinking he would not go far, and knowing he could hear or smell nothing in that rush of rain and crashing of thunder, I crept slowly after him. There he was, hunched up in the lee of a big hemlock, ears drooping, legs streaming, and little spurts of mist popping up from his pelted hide. Though woebegone enough, he had not forgotten caution; oh no! trust an old buck for that in any weather. His tail was to the tree, his head turned warily to the trail over which he had come. And there I left him, wishing as I turned back that he would let me stand under his hemlock, or else come and share my fir, just for a little company.

Near the lower end of my pond was still another tree which I must revisit; yes, surely, not only for its happy memories, but also in anticipation of some merry surprise, of which it seemed to have endless store. It stood on a bank overlooking a sunny dell in the woods, a wonderfully pleasant place where no wind entered, where the air was always fragrant, and a runlet of cool water sang a little tune to itself all day long. Its gnarled trunk was scarcely more than a shell, which boomed like a drum when a woodpecker sounded it; and above were hollow limbs with knot-hole entrances, offering hospitality to any wild creature in search of a weather-proof den or nesting-place.

The first time I passed this old tree a family of red squirrels were laying claim to it in a tiff with some larger beast or bird, which slipped away as I approached. The next time I saw it, a year later, it was silent and apparently deserted; but as I rose from drinking at the runlet the head of a little gray owl appeared at a knot-hole. For full ten minutes he remained there motionless, without word or sign or even a blink to say that he was watching me, though it was undoubtedly some noise or stir of mine which brought him up to his window.

After that I fell in the way of turning aside to loaf awhile under the inn-tree; and rarely could one loaf there very long without overhearing something not intended for a stranger's ear, some low dialogue or hammering signal or petulant whining or cautious scratching, to remind one of the running comedy of the woods. It was evidently an exchange, a crossroad or meeting-place for the wood folk, calling in every passer-by as a certain store or corner of a sleepy town invites all idlers, boys and stray dogs, while other stores or corners are empty, save for women folk, and quite respectable.

Once in the late morning, as I sat with an ear to the resonant shell, listening to the talk of unseen creatures which I fancied were young 'coons, a big log-cock flashed into the old tree, drew himself up on a stub over my head, and seemed to cock his ear at the voices to which I had been listening.

Now the log-cock is naturally a wary bird, shy and difficult of approach; but this gorgeous fellow with the scarlet crest became almost sociable in his curiosity, perhaps because the place was so quiet, so friendly, with no motion or hint of danger to disturb its tranquillity. He saw me at once, as the change in his bright eye plainly said; but, deceived by my stillness or the sober coloring of my clothes, he set me down as a tree-fungus or mushroom that had grown since his last visit, and looked about for something more interesting. When I called his attention by a curt nod, telling him that this was no dull mushroom, he came down at once to light against the side of the tree, where he examined my head minutely. Learning nothing from my wink, he went around the tree in a series of side-jumps to have a look from the other side; then he hopped up and down, this side or that, all the while uttering a low surprised chatter. Even when I began to flip bits of wood at him (for he soon grew impatient, and interrupted the 'coon talk by an unseemly rapping), instead of rushing off in alarm, he twice followed a missile that rattled near him, as if to demand, "Well, what in the world sent *you* flying?" Presently he sent forth a call, not the loud, high, prolonged note which you hear from him at a distance, but a soft, wheedling *ah-koo! ah-koo!* only twice repeated. When his call was answered in a different strain, a questioning strain it seemed to me, he darted away and returned within the minute accompanied by another log-cock.

But enough of such pictures! They flash joyously upon the mental vision whenever one recalls a cherished spot in the woods, but fade quickly if one attempts to hold or describe them, saying as they vanish that the lure of solitary lakes, the companionship of trees, the fascination of wild creatures that hide and look forth with roundly curious eyes at a stranger's approach,—these are matters that can never be set down in words: the best always escapes in the telling. I meant only to say (when my pine lifted its crown in the light of an evening sky, and then the mink family came dodging along the shore of memory, and the buck and the log-cock interrupted to urge me be sure and tell the happiest part of the story before I made an end) that many pleasant memories greeted me as I came down the silent trail after a long absence. In the distance sounded a lusty quacking; my imagination painted the mallards at the end of the alder run, with sunshiny water and crimson bog and misty-green larches around them, as a frame for the picture; and then the whole beautiful anticipation came tumbling in ruin about my ears.

Before I reached my pond, before I saw the welcoming gleam of it even, I was at every step going over my shoetops in water, where formerly I had always found dry footing. Something disastrous had happened in my absence; the whole bog was overflowed; around it was no mist of delicate foliage but only skeleton trees, stark and pitiful. In my heart I was berating the lumbermen, whose ugly works are the ruination of every place they visit, when at last I waded to an opening that gave outlook on my pond; and the first thing I noticed, as my eyes swept the familiar scene, was a beaver-house cocked up on the shore, like a warning sign of new ownership.

It is true that blessings brighten as they take their flight: not till I read that crude sign of dispossession did I know how much pleasure my little pond had given me. The lonely beauty which could quiet a man like a psalm, or like an Indian's wordless prayer; the glimpses of wild creatures at home and unafraid; the succession of radiant pictures, at sunny midday, or beneath the hushed twilight, or in the expectant morning before the shadows come,—all these had suddenly taken wing, driven away by mud-grubbing animals with a notion in their dull heads that they wanted deeper water about the site they had chosen for their house of sticks. It was too bad, too hopeless! I might have prevented the ruin had I known; but now it was beyond all remedy. With a different interest, therefore, and still resentful that my pond was spoiled as thoroughly as any lumberman would have spoiled it, I made my way around the flood to examine the beavers' work at the outlet.

Beaver Work

XV

HIDDEN among the larches at the lower end of my pond was a tiny outgoing stream, which had proved hard to find when first I explored the region, and almost impossible to follow afterward. Under a fallen log, so weathered and mossy that it seemed part of the natural shore, a volume of water escaped without ripple or murmur, wandering away under bending grasses to lose itself in an alder swamp, where innumerable channels offered it lingering passage. From the swamp it found its way, creepingly, among brooding cedars to a little brook, which went singing far down through the woods to Upper Pine Pond; and beyond that on the farther side was a long deadwater, and then Pine Stream making its tortuous way through an untraveled region to the Penobscot. The nearest beavers, a colony of four lodges which I unearthed on a hidden branch of Pine Stream, were twelve or fifteen miles away, as the water flowed; yet over all that distance an exploring family had made its lonely way, guided at every turn by the flavor of distant springs, till one after another they crept under the fallen log and entered my pond, which was solitary enough to satisfy even their pioneer instincts. They had first picked a site for their new lodge, on a point overlooking the lower half of the pond, and had then gone back to the outlet to raise the water.

Their dam was a rare piece of wild engineering; so much I had to confess, even while I wished that the beavers had chosen some other place to display their craft. Finding where the water escaped, they stopped the opening beneath the log, and made a bank of mud and alder-brush above it. This bank was carried out a dozen feet or more on either side of the stream, the ends being bent forward (toward the pond above) so as to make a very fine concave arch. On a small or quiet stream like this, beavers almost invariably build a straight dam; and where swift water calls for a stronger or curving structure, they present the convex side to the current; but here they had reversed both rules, for some reason or impulse which I could not fathom,— except on the improbable assumption that the animals could foresee the end of their work from the beginning. The finished dam was an amazingly good one, as you shall see; but whether it resulted from planning or happy experiment or just following the water, only a certain old beaver could tell.

Since there was no other outlet to my pond, the beavers were obliged to build here; but the site was a poor one, the land being uniformly low on all sides, and no sooner did they finish their dam than the rising water flowed around both ends of it. To remedy this they pushed out a curving wing from either end of their first arch, so that the line of their dam was now a pretty triple-curve. Again and again the outgoing water crept around the obstacle; each time the beavers added other curving wings, now on this side, now on that, bending them steadily forward till the top of their dam suggested the rim of an enormous scallop-shell. Then, finding the water deep enough for their needs, they thrust out a straight wing from either end of their dam, resting their work on the slopes of two hillocks in the woods, some fifty yards apart,—this in a straight line, or across the hinge of the scallop-shell: if measured on the curves, their dam was three or four times that length. Their next task was to build a lodge on the point above; then they dug a canal through the bog to the nearest grove of hardwood, and cut down a liberal part of the trees for their winter supply of bark. The branches of these trees had been cut into convenient lengths, floated through the canal, and stored in a great food-pile in the deep water near the lodge.

When I found the dam, several deer (to judge from the tracks) were already using the top of it as a runway in passing from the flooded ground on one side of the pond to the other. From either end a game-trail led upward along the shore, no longer following immemorial paths over the bog, which was submerged with all its splendor of color, but making a new and rougher way through the black growth. When I followed one of these trails it led me completely around the pond, going confidently till it neared the salt-lick, where it halted, wavered and trickled out in aimless wanderings. There, where once the ground was trodden smooth by many feet, was now no ground to be seen. The precious spring, over which a thousand generations of deer had lingered, had vanished in a dull waste of water. Twice I watched the place from early morning till owls began to cry the twilight; in that time only a few animals appeared, singly, at long intervals; and after wandering about as if seeking something and finding it not, they disappeared in the dusky woods.

And so I went away, looking for the last time sadly on the little pond, as upon a place one has owned and loved, but which has passed into other hands. Though the wild ducks still breed or gather there, it is no longer the same. There is no restful spot from which to watch the waters dance with the wind, or frown at the cloud, or smile at the sunshine; the little larches are all dead beside their ancestors; the carpet of colored moss is but a memory. When the beavers go to pioneer a remoter spot, I shall break their dam and let the water return to its ancient level. Then, if happily I live long enough for another fringe of larches to grow, and another mossy rug to crimson under the waning sun, perhaps it will be my pond once more.

THE END

A CATALOG OF SELECTED DOVER
BOOKS IN ALL FIELDS OF INTEREST

CONCERNING THE SPIRITUAL IN ART, Wassily Kandinsky. Pioneering work by father of abstract art. Thoughts on color theory, nature of art. Analysis of earlier masters. 12 illustrations. 80pp. of text. 5³/₈ x 8¹/₂.   0-486-23411-8

CELTIC ART: The Methods of Construction, George Bain. Simple geometric techniques for making Celtic interlacements, spirals, Kells-type initials, animals, humans, etc. Over 500 illustrations. 160pp. 9 x 12. (Available in U.S. only.)   0-486-22923-8

AN ATLAS OF ANATOMY FOR ARTISTS, Fritz Schider. Most thorough reference work on art anatomy in the world. Hundreds of illustrations, including selections from works by Vesalius, Leonardo, Goya, Ingres, Michelangelo, others. 593 illustrations. 192pp. 7¹/₈ x 10¹/₄.   0-486-20241-0

CELTIC HAND STROKE-BY-STROKE (Irish Half-Uncial from "The Book of Kells"): An Arthur Baker Calligraphy Manual, Arthur Baker. Complete guide to creating each letter of the alphabet in distinctive Celtic manner. Covers hand position, strokes, pens, inks, paper, more. Illustrated. 48pp. 8¹/₄ x 11.   0-486-24336-2

EASY ORIGAMI, John Montroll. Charming collection of 32 projects (hat, cup, pelican, piano, swan, many more) specially designed for the novice origami hobbyist. Clearly illustrated easy-to-follow instructions insure that even beginning papercrafters will achieve successful results. 48pp. 8¹/₄ x 11.   0-486-27298-2

BLOOMINGDALE'S ILLUSTRATED 1886 CATALOG: Fashions, Dry Goods and Housewares, Bloomingdale Brothers. Famed merchants' extremely rare catalog depicting about 1,700 products: clothing, housewares, firearms, dry goods, jewelry, more. Invaluable for dating, identifying vintage items. Also, copyright-free graphics for artists, designers. Co-published with Henry Ford Museum & Greenfield Village. 160pp. 8¹/₄ x 11.

0-486-25780-0

THE ART OF WORLDLY WISDOM, Baltasar Gracian. "Think with the few and speak with the many," "Friends are a second existence," and "Be able to forget" are among this 1637 volume's 300 pithy maxims. A perfect source of mental and spiritual refreshment, it can be opened at random and appreciated either in brief or at length. 128pp. 5³/₈ x 8¹/₂.

0-486-44034-6

JOHNSON'S DICTIONARY: A Modern Selection, Samuel Johnson (E. L. McAdam and George Milne, eds.). This modern version reduces the original 1755 edition's 2,300 pages of definitions and literary examples to a more manageable length, retaining the verbal pleasure and historical curiosity of the original. 480pp. 5³/₁₆ x 8¹/₄.

0-486-44089-3

ADVENTURES OF HUCKLEBERRY FINN, Mark Twain, Illustrated by E. W. Kemble. A work of eternal richness and complexity, a source of ongoing critical debate, and a literary landmark, Twain's 1885 masterpiece about a barefoot boy's journey of self-discovery has enthralled readers around the world. This handsome clothbound reproduction of the first edition features all 174 of the original black-and-white illustrations. 368pp. 5³/₈ x 8¹/₂.

0-486-44322-1

STICKLEY CRAFTSMAN FURNITURE CATALOGS, Gustav Stickley and L. & J. G. Stickley. Beautiful, functional furniture in two authentic catalogs from 1910. 594 illustrations, including 277 photos, show settles, rockers, armchairs, reclining chairs, bookcases, desks, tables. 183pp. 6¹/₂ x 9¹/₄. 0-486-23838-5

AMERICAN LOCOMOTIVES IN HISTORIC PHOTOGRAPHS: 1858 to 1949, Ron Ziel (ed.). A rare collection of 126 meticulously detailed official photographs, called "builder portraits," of American locomotives that majestically chronicle the rise of steam locomotive power in America. Introduction. Detailed captions. xi+ 129pp. 9 x 12. 0-486-27393-8

AMERICA'S LIGHTHOUSES: An Illustrated History, Francis Ross Holland, Jr. Delightfully written, profusely illustrated fact-filled survey of over 200 American lighthouses since 1716. History, anecdotes, technological advances, more. 240pp. 8 x 10³/₄. 0-486-25576-X

TOWARDS A NEW ARCHITECTURE, Le Corbusier. Pioneering manifesto by founder of "International School." Technical and aesthetic theories, views of industry, economics, relation of form to function, "mass-production split" and much more. Profusely illustrated. 320pp. 6¹/₈ x 9¹/₄. (Available in U.S. only.) 0-486-25023-7

HOW THE OTHER HALF LIVES, Jacob Riis. Famous journalistic record, exposing poverty and degradation of New York slums around 1900, by major social reformer. 100 striking and influential photographs. 233pp. 10 x 7⁷/₈. 0-486-22012-5

FRUIT KEY AND TWIG KEY TO TREES AND SHRUBS, William M. Harlow. One of the handiest and most widely used identification aids. Fruit key covers 120 deciduous and evergreen species; twig key 160 deciduous species. Easily used. Over 300 photographs. 126pp. 5³/₈ x 8¹/₂. 0-486-20511-8

COMMON BIRD SONGS, Dr. Donald J. Borror. Songs of 60 most common U.S. birds: robins, sparrows, cardinals, bluejays, finches, more—arranged in order of increasing complexity. Up to 9 variations of songs of each species.
Cassette and manual 0-486-99911-4

ORCHIDS AS HOUSE PLANTS, Rebecca Tyson Northen. Grow cattleyas and many other kinds of orchids—in a window, in a case, or under artificial light. 63 illustrations. 148pp. 5³/₈ x 8¹/₂. 0-486-23261-1

MONSTER MAZES, Dave Phillips. Masterful mazes at four levels of difficulty. Avoid deadly perils and evil creatures to find magical treasures. Solutions for all 32 exciting illustrated puzzles. 48pp. 8¹/₄ x 11. 0-486-26005-4

MOZART'S DON GIOVANNI (DOVER OPERA LIBRETTO SERIES), Wolfgang Amadeus Mozart. Introduced and translated by Ellen H. Bleiler. Standard Italian libretto, with complete English translation. Convenient and thoroughly portable—an ideal companion for reading along with a recording or the performance itself. Introduction. List of characters. Plot summary. 121pp. 5¹/₄ x 8¹/₂. 0-486-24944-1

FRANK LLOYD WRIGHT'S DANA HOUSE, Donald Hoffmann. Pictorial essay of residential masterpiece with over 160 interior and exterior photos, plans, elevations, sketches and studies. 128pp. 9¹/₄ x 10³/₄. 0-486-29120-0

HOW TO DO BEADWORK, Mary White. Fundamental book on craft from simple projects to five-bead chains and woven works. 106 illustrations. 142pp. 5⅜ x 8.
0-486-20697-1

THE 1912 AND 1915 GUSTAV STICKLEY FURNITURE CATALOGS, Gustav Stickley. With over 200 detailed illustrations and descriptions, these two catalogs are essential reading and reference materials and identification guides for Stickley furniture. Captions cite materials, dimensions and prices. 112pp. 6½ x 9¼.   0-486-26676-1

SIX GREAT DIALOGUES: Apology, Crito, Phaedo, Phaedrus, Symposium, The Republic, Plato, translated by Benjamin Jowett. Plato's Dialogues rank among Western civilization's most important and influential philosophical works. These 6 selections of his major works explore a broad range of enduringly relevant issues. Authoritative Jowett translations. 480pp. 5³⁄₁₆ x 8¼.   0-486-45465-7

DEMONOLATRY: An Account of the Historical Practice of Witchcraft, Nicolas Remy, edited with an Introduction and Notes by Montague Summers, translated by E. A. Ashwin. This extremely influential 1595 study was frequently cited at witchcraft trials. In addition to lurid details of satanic pacts and sexual perversity, it presents the particulars of numerous court cases. 240pp. 6½ x 9¼.   0-486-46137-8

VICTORIAN FASHIONS AND COSTUMES FROM HARPER'S BAZAAR, 1867–1898, Stella Blum (ed.). Day costumes, evening wear, sports clothes, shoes, hats, other accessories in over 1,000 detailed engravings. 320pp. 9⅜ x 12¼.   0-486-22990-4

THE LONG ISLAND RAIL ROAD IN EARLY PHOTOGRAPHS, Ron Ziel. Over 220 rare photos, informative text document origin (1844) and development of rail service on Long Island. Vintage views of early trains, locomotives, stations, passengers, crews, much more. Captions. 8⅞ x 11¾.   0-486-26301-0

VOYAGE OF THE LIBERDADE, Joshua Slocum. Great 19th-century mariner's thrilling, first-hand account of the wreck of his ship off South America, the 35-foot boat he built from the wreckage, and its remarkable voyage home. 128pp. 5⅜ x 8½.   0-486-40022-0

TEN BOOKS ON ARCHITECTURE, Vitruvius. The most important book ever written on architecture. Early Roman aesthetics, technology, classical orders, site selection, all other aspects. Morgan translation. 331pp. 5⅜ x 8½.   0-486-20645-9

THE HUMAN FIGURE IN MOTION, Eadweard Muybridge. More than 4,500 stopped-action photos, in action series, showing undraped men, women, children jumping, lying down, throwing, sitting, wrestling, carrying, etc. 390pp. 7⅞ x 10⅝.
0-486-20204-6 Clothbd.

TREES OF THE EASTERN AND CENTRAL UNITED STATES AND CANADA, William M. Harlow. Best one-volume guide to 140 trees. Full descriptions, woodlore, range, etc. Over 600 illustrations. Handy size. 288pp. 4½ x 6⅜.   0-486-20395-6

MY FIRST BOOK OF TCHAIKOVSKY: Favorite Pieces in Easy Piano Arrangements, edited by David Dutkanicz. These special arrangements of favorite Tchaikovsky themes are ideal for beginner pianists, child or adult. Contents include themes from "The Nutcracker," "March Slav," Symphonies Nos. 5 and 6, "Swan Lake," "Sleeping Beauty," and more. 48pp. 8¼ x 11.   0-486-46416-4

BIG BOOK OF MAZES AND LABYRINTHS, Walter Shepherd. 50 mazes and labyrinths in all—classical, solid, ripple, and more—in one great volume. Perfect inexpensive puzzler for clever youngsters. Full solutions. 112pp. 8⅛ x 11.   0-486-22951-3

PIANO TUNING, J. Cree Fischer. Clearest, best book for beginner, amateur. Simple repairs, raising dropped notes, tuning by easy method of flattened fifths. No previous skills needed. 4 illustrations. 201pp. 5⅜ x 8½.   0-486-23267-0

HINTS TO SINGERS, Lillian Nordica. Selecting the right teacher, developing confidence, overcoming stage fright, and many other important skills receive thoughtful discussion in this indispensible guide, written by a world-famous diva of four decades' experience. 96pp. 5³⁄₈ x 8¹⁄₂. 0-486-40094-8

THE COMPLETE NONSENSE OF EDWARD LEAR, Edward Lear. All nonsense limericks, zany alphabets, Owl and Pussycat, songs, nonsense botany, etc., illustrated by Lear. Total of 320pp. 5³⁄₈ x 8¹⁄₂. (Available in U.S. only.) 0-486-20167-8

VICTORIAN PARLOUR POETRY: An Annotated Anthology, Michael R. Turner. 117 gems by Longfellow, Tennyson, Browning, many lesser-known poets. "The Village Blacksmith," "Curfew Must Not Ring Tonight," "Only a Baby Small," dozens more, often difficult to find elsewhere. Index of poets, titles, first lines. xxiii + 325pp. 5³⁄₈ x 8¹⁄₄.
0-486-27044-0

DUBLINERS, James Joyce. Fifteen stories offer vivid, tightly focused observations of the lives of Dublin's poorer classes. At least one, "The Dead," is considered a masterpiece. Reprinted complete and unabridged from standard edition. 160pp. 5³⁄₁₆ x 8¹⁄₄.
0-486-26870-5

THE LITTLE RED SCHOOLHOUSE, Eric Sloane. Harkening back to a time when the three Rs stood for reading, 'riting, and religion, Sloane's sketchbook explores the history of early American schools. Includes marvelous illustrations of one-room New England schoolhouses, desks, and benches. 48pp. 8¹⁄₄ x 11. 0-486-45604-8

THE BOOK OF THE SACRED MAGIC OF ABRAMELIN THE MAGE, translated by S. MacGregor Mathers. Medieval manuscript of ceremonial magic. Basic document in Aleister Crowley, Golden Dawn groups. 268pp. 5³⁄₈ x 8¹⁄₂. 0-486-23211-5

THE BATTLES THAT CHANGED HISTORY, Fletcher Pratt. Eminent historian profiles 16 crucial conflicts, ancient to modern, that changed the course of civilization. 352pp. 5³⁄₈ x 8¹⁄₂. 0-486-41129-X

NEW RUSSIAN-ENGLISH AND ENGLISH-RUSSIAN DICTIONARY, M. A. O'Brien. This is a remarkably handy Russian dictionary, containing a surprising amount of information, including over 70,000 entries. 366pp. 4¹⁄₂ x 6¹⁄₈. 0-486-20208-9

NEW YORK IN THE FORTIES, Andreas Feininger. 162 brilliant photographs by the well-known photographer, formerly with *Life* magazine. Commuters, shoppers, Times Square at night, much else from city at its peak. Captions by John von Hartz. 181pp. 9¹⁄₄ x 10³⁄₄. 0-486-23585-8

INDIAN SIGN LANGUAGE, William Tomkins. Over 525 signs developed by Sioux and other tribes. Written instructions and diagrams. Also 290 pictographs. 111pp. 6¹⁄₈ x 9¹⁄₄.
0-486-22029-X

ANATOMY: A Complete Guide for Artists, Joseph Sheppard. A master of figure drawing shows artists how to render human anatomy convincingly. Over 460 illustrations. 224pp. 8³⁄₈ x 11¹⁄₄. 0-486-27279-6

MEDIEVAL CALLIGRAPHY: Its History and Technique, Marc Drogin. Spirited history, comprehensive instruction manual covers 13 styles (ca. 4th century through 15th). Excellent photographs; directions for duplicating medieval techniques with modern tools. 224pp. 8³⁄₈ x 11¹⁄₄. 0-486-26142-5

CATALOG OF DOVER BOOKS

LIGHT AND SHADE: A Classic Approach to Three-Dimensional Drawing, Mrs. Mary P. Merrifield. Handy reference clearly demonstrates principles of light and shade by revealing effects of common daylight, sunshine, and candle or artificial light on geometrical solids. 13 plates. 64pp. 5³/₈ x 8¹/₂. 0-486-44143-1

ASTROLOGY AND ASTRONOMY: A Pictorial Archive of Signs and Symbols, Ernst and Johanna Lehner. Treasure trove of stories, lore, and myth, accompanied by more than 300 rare illustrations of planets, the Milky Way, signs of the zodiac, comets, meteors, and other astronomical phenomena. 192pp. 8³/₈ x 11. 0-486-43981-X

JEWELRY MAKING: Techniques for Metal, Tim McCreight. Easy-to-follow instructions and carefully executed illustrations describe tools and techniques, use of gems and enamels, wire inlay, casting, and other topics. 72 line illustrations and diagrams. 176pp. 8¹/₄ x 10⁷/₈. 0-486-44043-5

MAKING BIRDHOUSES: Easy and Advanced Projects, Gladstone Califf. Easy-to-follow instructions include diagrams for everything from a one-room house for bluebirds to a forty-two-room structure for purple martins. 56 plates; 4 figures. 80pp. 8³/₄ x 6⁵/₈.
0-486-44183-0

LITTLE BOOK OF LOG CABINS: How to Build and Furnish Them, William S. Wicks. Handy how-to manual, with instructions and illustrations for building cabins in the Adirondack style, fireplaces, stairways, furniture, beamed ceilings, and more. 102 line drawings. 96pp. 8³/₄ x 6⁵/₈. 0-486-44259-4

THE SEASONS OF AMERICA PAST, Eric Sloane. From "sugaring time" and strawberry picking to Indian summer and fall harvest, a whole year's activities described in charming prose and enhanced with 79 of the author's own illustrations. 160pp. 8¹/₄ x 11.
0-486-44220-9

THE METROPOLIS OF TOMORROW, Hugh Ferriss. Generous, prophetic vision of the metropolis of the future, as perceived in 1929. Powerful illustrations of towering structures, wide avenues, and rooftop parks—all features in many of today's modern cities. 59 illustrations. 144pp. 8¹/₄ x 11. 0-486-43727-2

THE PATH TO ROME, Hilaire Belloc. This 1902 memoir abounds in lively vignettes from a vanished time, recounting a pilgrimage on foot across the Alps and Apennines in order to "see all Europe which the Christian Faith has saved." 77 of the author's original line drawings complement his sparkling prose. 272pp. 5³/₈ x 8¹/₂. 0-486-44001-X

THE HISTORY OF RASSELAS: Prince of Abissinia, Samuel Johnson. Distinguished English writer attacks eighteenth-century optimism and man's unrealistic estimates of what life has to offer. 112pp. 5³/₈ x 8¹/₂. 0-486-44094-X

A VOYAGE TO ARCTURUS, David Lindsay. A brilliant flight of pure fancy, where wild creatures crowd the fantastic landscape and demented torturers dominate victims with their bizarre mental powers. 272pp. 5³/₈ x 8¹/₂. 0-486-44198-9

Paperbound unless otherwise indicated. Available at your book dealer, online at **www.doverpublications.com**, or by writing to Dept. GI, Dover Publications, Inc., 31 East 2nd Street, Mineola, NY 11501. For current price information or for free catalogs (please indicate field of interest), write to Dover Publications or log on to **www.doverpublications.com** and see every Dover book in print. Dover publishes more than 400 books each year on science, elementary and advanced mathematics, biology, music, art, literary history, social sciences, and other areas.